# COMPUTER VISION
## From Surfaces to 3D Objects

# COMPUTER VISION
## From Surfaces to 3D Objects

**Edited by**

# Christopher W. Tyler

Smith-Kettlewell Eye Research Institute
San Francisco, California, USA

CRC Press
Taylor & Francis Group
Boca Raton London New York

CRC Press is an imprint of the
Taylor & Francis Group, an **informa** business
A CHAPMAN & HALL BOOK

CRC Press
Taylor & Francis Group
6000 Broken Sound Parkway NW, Suite 300
Boca Raton, FL 33487-2742

First issued in paperback 2019

© 2011 by Taylor & Francis Group, LLC
CRC Press is an imprint of Taylor & Francis Group, an Informa business

No claim to original U.S. Government works

ISBN-13: 978-1-4398-1712-4 (hbk)
ISBN-13: 978-0-367-38309-1 (pbk)

# Table of Contents

# Introduction: The Role of Midlevel Surface Representation in 3D Object Encoding

Christopher W. Tyler

## VISUAL ENCODING

The primary goal of visual encoding is to determine the nature and motion of the objects in the surrounding environment. In order to plan and coordinate actions, we need a functional representation of the dynamics of the scene layout and of the spatial configuration and the dynamics of the objects within it. The properties of the visual array, however, have a metric structure entirely different from that of the spatial configuration of the objects. Physically, objects consist of aggregates of particles that cohere together. Objects may be rigid or flexible but, in either case, an invariant set of particles is connected to form a given object.

Although the objects may be stable and invariant in the scene before us, the cues that convey the presence of the objects to the eyes are much less stable. They may change in luminance or color, they may be disrupted by reflections or highlights or occlusion by intervening objects. The various cues carrying the information about the physical object structure, such as edge structure, binocular disparity, color, shading, texture, and motion vector fields, typically carry information that is inconsistent. Many of these cues may be sparse, with missing information about the object structure across gaps where there are no edge or texture cues to carry information about the object shape (see Figure 0.1).

FIGURE 0.1    Example of a scene with many objects viewed at oblique angles, requiring advanced three-dimensional surface reconstruction for a proper understanding of the object structure. Although the objects are well-defined by luminance and color transitions, many of the objects have minimal surface texture between the edges. The object shape cannot be determined from the two-dimensional shape of the outline. The three-dimensional structure of the objects requires interpolation from sparse depth cues across regions with no shape information. (From TurboPhoto. With permission.)

The typical approach to computational object understanding is to derive the shape from the two-dimensional (2D) outline of the objects. For complex object structures, however, object shape cannot be determined from the 2D shape of the outline. In the scene depicted in Figure 0.1, many of the objects have nontraditional shapes that have minimal surface texture between the edges and are set at oblique angles to the camera plane. Although the structural edges are well-defined by luminance and color transitions, there are shadows that have to be understood as nonstructural edges, and the structural edges have to be encoded in terms of their three-dimensional (3D) spatial configuration, not just their 2D location. In natural scenes, the 3D information is provided by stereoscopic disparity, linear perspective, motion parallax, and other cues. In the 2D representation of Figure 1, of course, only linear perspective is available, and in a rather limited fashion in the absence of line or grid texture on the surfaces.

Consequently, the 3D structure of the objects requires both

1. Global integration among the various edge cues to derive the best estimate of the edge structure in 3D space, and

2. 3D interpolation from sparse depth cues across regions with no shape information

These are challenging tasks involving an explicit 3D representation of the visual scene, which have not been addressed to be either computational techniques or empirical studies of how the visual system achieves them.

Thus, a primary requirement of neural or computational representations of the structure of objects is the filling-in of details of the three-dimensional object structure across regions of missing or discrepant information in the local visual cues. Generating such three-dimensional representations is a fundamental role of the posterior brain regions, but the cortical architecture for the joint processing of multiple depth cues is poorly understood in the brain, especially in terms of the joint processing of diverse visual cues. Computational approaches to the issue of the structure of objects tend to take either a low-level or a high-level approach to the problem. Low-level approaches begin with local feature recognition and an attempt to build up the object representation by hierarchical convergence, using primarily feedforward logic with some recurrent feedback tuning of the results (Marr, 1982; Grossberg, Kuhlmann, and Mingolla, 2007). High-level, or Bayesian, approaches begin with the vocabulary of likely object structures and look for evidence in the visual array as to which object might be there (Huang and Russell, 1998; Rue and Hurn, 1999; Moghaddam, 2001; Stormont, 2007). Both approaches work much better for objects with a stable 2D structure (when translating through the visual scene) than for manipulation of objects with a full 3D structure, such as those in Figure 1, or for locomotion through a complex 3D scene.

## SURFACES AS A MID-LEVEL INVARIANT IN VISUAL ENCODING

A more fruitful approach to the issue of 3D object structure is to focus the analysis on midlevel invariants to the object structure, such as surfaces, symmetry, rigidity, texture invariants, or surface reflectance properties.

Each of these properties is invariant under transformations of 3D pose, viewpoint, illumination level, haze contrast, and other variations of environmental conditions. In mathematical terminology, given a set of object points $X$ with an equivalence relation A ~ A′ on it, any function $f : X \rightarrow Y$ is constant over the equivalence classes of the transformations. Various computational analyses have incorporated such invariants in their object-recognition schemes, but a neglected aspect of midlevel vision is the 3D surface structure that is an inescapable property of objects in the world. The primary topic of this book is therefore to focus attention on this important midlevel invariant of object representation, both to determine its role in human vision and to analyze the potential of surface structure as a computational representation to improve the capability of decoding the object structure of the environment.

Surfaces are a key property of our interaction with objects in the world. It is very unusual to experience objects, either tactilely or visually, except through their surfaces. Even transparent objects are experienced in relation to their surfaces, with the material between the surfaces being invisible by virtue of the transparency. The difference between transparent and opaque surfaces is that only the near surfaces of opaque objects are visible, while in transparent objects both the near and far surfaces are visible. (Only translucent objects are experienced in an interior sense, as the light passes through them to illuminate the density of the material. Nevertheless, the interior is featureless and has no appreciable shape information.) Thus, the surfaces predominate for virtually all objects. Developing a means of representing the proliferation of surfaces before us is therefore a key stage in the processing of objects.

For planar surfaces, extended cues such as luminance shading, linear perspective, aspect ratio of square or round objects, and texture gradient can each specify the slant of a planar surface (see Figure 0.2). Zimmerman, Legge, and Cavanagh (1995) performed experiments to measure the accuracy of surface slant from judgments of the relative lengths of a pair of orthogonal lines embedded in one surface of a full visual scene. Slant judgments are accurate to within 3° for all three cue types, with no evidence of the recession to the frontal plane expected if the pictorial surface was contaminating the estimations. Depth estimates of disconnected surfaces were, however, strongly compressed. Such results emphasize the key role of surface reconstruction in human depth estimation.

In assessing the placement of surfaces in space, Alhazen, in his *Optics* (AD ~1000), argued that "sight does not perceive the magnitudes of distances of visible objects from itself unless these distances extend along a

FIGURE 0.2    Scene in which texture gradients help to determine the shapes of the objects. (From http://www.freephotosbank.com/5065.html.)

series of continuous bodies, and unless sight perceives those bodies and their magnitudes" (Sabra, 1989, 155). He supported his argument with examples in which the distances are correctly estimated for objects seen resting on a continuous ground, and misperceived as close to each other when the view of the ground is obstructed. Alhazen's explanation of distance perception was arrived at independently by Gibson (1950), who called it the "ground theory" of space perception, based on the same idea that the visual system uses the ground surface as a reference frame for coding location in space. The orientation of the ground plane relative to the observer is not a given, but it has to be derived from stereoscopic, perspective, and texture gradient cues. Once established, it can form a basis not only for distance estimation but for estimation of the relative orientation of other surfaces in the scene.

## THE ANALYSIS OF SURFACE CURVATURE

When surfaces are completed in three dimensions, not just in two dimensions, the defining property is the curvature at each point in the surface. Curvature is a subtle property defined in many ways, but the three most notable properties of surface curvature are its *local curvature* $\kappa$, its *mean curvature* $H$, and its *Gaussian curvature* $K$. These quantities are all continuous properties of the surface.

At each point in the surface, a one-dimensional (1D) cut through the surface defines the local curvature $\kappa$ (see Figure 0.3), which is most easily conceptualized as the reciprocal of the radius of curvature for the circle tangent to the curve of the surface at this point. For a cut at some angle $\theta$ in the tangent plane:

$$\kappa_\theta = \frac{1}{r_\theta} \tag{0.1}$$

As $\theta$ rotates around the point, it describes a curvature function for the surface. Any point on a smooth surface has only one maximum and one minimum direction of curvature, and the directions of maximum and minimum curvature are always at right angles to each other in the tangent plane. Gauss defined these extrema as the *principal curvatures* for that point in the surface:

$$\kappa_1 = \max(\kappa_\theta) \quad \kappa_2 = \min(\kappa_\theta) \tag{0.2}$$

where

$$\theta_{\kappa_2} = \theta_{\kappa_1} \pm \frac{\pi}{2}$$

The principal curvatures at every point are the fundamental intrinsic properties of the surface. The intrinsic 2D curvature is now known as the Gaussian curvature, $K$, given by the product of the two principal curvatures:

$$K = \kappa_1 \cdot \kappa_2 \tag{0.3}$$

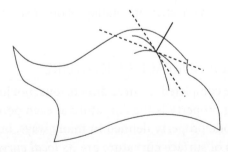

FIGURE 0.3 Surface curvature. At a local point on any surface, there are orthogonal maximum and minimum curvatures whose tangents (dashed lines) define a local coordinate frame on which the surface normal (thick line) is erected.

If a surface is intrinsically flat, like a sheet of paper, it can never be curved to a form with intrinsic curvature, such as the surface of a sphere. Even if the flat sheet is curved to give the curvature in one direction a positive value, the curvature in the other (minimal) direction will be zero (like a cylinder). Hence, the product forming the Gaussian curvature (Equation 0.3) will remain zero. For any point of a sphere (or any other convex form), however, both the maximum and the minimum curvatures will have the same sign, and $K$ will be positive everywhere on the surface.

When Gaussian curvature is applied to objects, one issue that arises is that the value of $K$ varies as an object changes in size. As a measure of shape, this is an undesirable property, because we think of a sphere, for example, as having the same shape regardless of its size. One option that may be suggested to overcome this problem is to normalize the local Gaussian curvature $K$ to the global mean curvature of the surface of an object $\hat{H}$ (see Equation 0.4) to provide a measure of the normalized Gaussian curvature $K_{norm}$. Objects meaningfully exist only when the mean curvature $\hat{H} > 0$, so this should be considered as a restriction on the concept of normalized Gaussian curvature (i.e., an isolated patch of surface with $\hat{H} <= 0$) is solely a mathematical construct that cannot be a real object). Normalized Gaussian curvature is thus expressed in polar angle coordinates with respect to the center of the object, $\Theta$, and is defined as

$$K_{norm}(\Theta) = K(\Theta)/\hat{H}, \quad \hat{H} > 0 \qquad (0.4)$$

where the mean curvature $H = (\kappa_1 + \kappa_2)/2$, and global mean curvature is the integral of the local mean curvature over all angles with respect to its center $\hat{H} = \int H\, d\Theta$.

Thus, $K_{norm} = 1$ for spheres of all radii, rather than varying with radius as does $K$, forming a map in spherical coordinates that effectively encodes the shape of the object with a function that is independent of its size. Regions where the local Gaussian curvature matches the global mean Gaussian curvature will have $K_{norm} = 1$, and regions that are flat like the facets of a crystal will have $K_{norm} = 0$, with high values where they transition from one facet to the next.

Another shape property of interest is mean curvature, defined as the mean of the two principal curvatures. Constant mean curvature specifies a class of surfaces in which $H$ is constant everywhere. In the limiting case, the mean curvature is zero everywhere, $H = 0$. This is the tendency exhibited by surfaces governed by the self-organizing principle of molecular surface tension, like soap films. The surface tension within the film operates to minimize the curvature of the surface in a manner that is defined by the perimeter to which the soap bubble clings. In principle, the shapes adopted by soap films on a wire frame are minimal surfaces. Note that if $H = 0$, then $\kappa_1 = -\kappa_2$ and the Gaussian curvature product is negative (or zero) throughout. Thus, minimal surfaces always have negative Gaussian curvature, meaning that they are locally saddle shaped with no rounded bulges or dimples of positive curvature. The surface tension principle operates to flatten any regions of net positive curvature until the mean curvature is zero.

Until 1967, it was thought that there were only two minimal surfaces exhibiting the property of $H = 0$, the negative-curvature hypersphere and the imaginary-curvature hyperboloid. Since that time, a large class of such surfaces has been discovered. These are very interesting surfaces with quasi-biological resonances, suggesting that the minimal surface principle is commonly adopted (or approximated) in biological systems (Figure 0.4). Thus, the simple specification of $H = 0$ generates a universe of interesting surface forms that may be of relevance in neural processing as much as they are in biological form. The ubiquitous operation of lateral inhibition at all levels in the nervous system is a minimal principle comparable with surface tension. It is generally viewed as operating on a 2D function of the neural array in some tissue such as the retina or visual cortex. However, for any form of neural connectivity that is effectively 3D, lateral inhibition may well operate to form minimal surfaces of neural connectivity with conforming mathematical minimal surface principle (or its dynamic manifestation).

A similar approach to this minimization for the 1D case over the domain $\Gamma(x)$ has been proposed by Nitzberg, Mumford, and Shiota (1993) as minima of the elastic functional on the curvature $\kappa$:

$$E = \int_{\Gamma} (v + \alpha\kappa^2)\, dx, \qquad (0.5)$$

where $v$ and $\alpha$ are constants.

Using a sparse set of depth cues $\Phi$, one may define a function $f: \mathbf{R}^3 \to \mathbf{R}$ that estimates the signed geometric distance to the unknown surface $S$, and then use a contouring algorithm to extract the surface function approximating its zero set:

$$Z(f) = \{\, S(\Phi) \in \mathbf{R}^3 : S(\Phi) = 0 \,\} \qquad (0.6)$$

To make the problem computationally tractable (and a realistic model for neural representation within a 2D cortical sheet), we may propose a

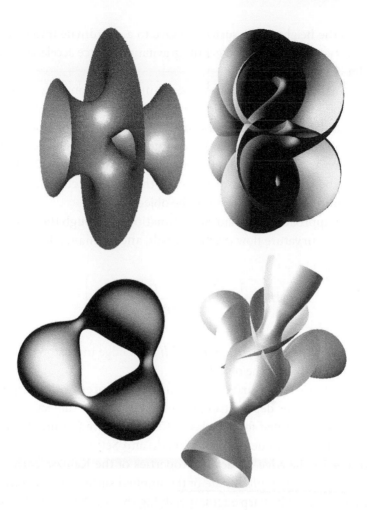

FIGURE 0.4 Examples of mathematical minimal surfaces. (Minimal surface images by Shoichi Fujimori, with permission.)

restriction to depth in the egocentric direction relative to the 2D array of the visual field, such that the surface is described by the function $S(x,y)$. This 2D case may be approached through the self-organizing principle that determines the shape of the soap-film surface is the minimization of the energy function defined according to

$$\min[E] = \min\left[\iint\left(k\sqrt{1+\left|\nabla h\right|^2} + \rho g h\right)\right] \quad (0.7)$$

where $h$ is the height of the surface relative to a coordinate frame, $\rho$ is the mass density per unit area, and $g$ is the gravitational force acceleration on the soap film.

These formulations, in terms of energy minimization, may provide the basis for the kind of optimization process that needs to be implemented in the neural processing network for the midlevel representation of the arbitrary object surfaces that are encountered in the visual scene. In the neural application, the constant $g$ can be a generalized to a 2D function that takes the role of the biasing effect of the distance cues that drive the minimization to match the form of the object surface.

Another approach to this surface estimation is through the evolution of Riemannian curvature flow (Sarti, Malladi, and Sethian, 2000):

$$H_g = \frac{1}{h^3} \vartheta_i \left( h \frac{\vartheta_i u}{\sqrt{\varepsilon + |\vartheta_1 u|^2 + |\vartheta_2 u|^2}} \right) \tag{0.8}$$

where $u(x,y)$ is the image function, and $h$ is an edge indicator function for the bounding edges of the object.

The feasibility of a surface reconstruction process being capable of generating accurate depth structure underlying the subjective contour generation is illustrated for the classic Kanizsa figure (Figure 0.5a) in the computational technique of Sarti, Malladi, and Sethian (2000). As shown in Figure 0.5b, the edge-attractant properties of the Kanizsa corners progressively convert the initial state of the implied surface into a convincing triangular mesa with sharp edges, specifying the *depth map* of the perceptual interpretation of the triangle in front of the pacmen disk segments. The resulting subjective surface is developed as a minimal surface with respect to distortions induced by the features in the image (Figure 0.5c). This computational morphogenesis of the surface reveals how the interactions within a continuous neural network could operate to generate the sharp subjective contours in the course of the 3D reconstruction of surfaces in the world.

Each of these methods incorporates some level of Bayesian inference as to the probable parameters of surface structure, but such inference may be applied with specific reference to the assessed probabilities of surfaces in the environment (Sullivan et al., 2001; Lee and Mumford, 2003; Yang and Purves, 2003). One goal of the book is to focus attention on such principles for the analysis of the properties of neural encoding of surfaces, or for the

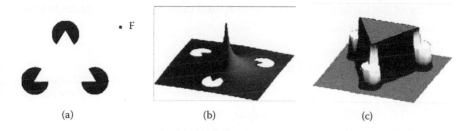

(a)     (b)     (c)

FIGURE 0.5   (a) The original Kanizsa (1976) illusory triangle. Importantly, no subjective contours are seen if the white region appears flat. Note also that the illusion disappears with fixation at point F to project the triangle to peripheral retina (Ramachandran et al., 1994). (b,c) Surface manifold output of the model of the perceptual surface reconstruction process by Sarti, Malladi, and Sethian (2000). Model of the perceptual surface reconstruction process. (b) Starting condition of the default surface superimposed on the figure. (c) Development of the surface toward the subjective surface. The original features are mapped in white.

neural encoding of other features of the visual world that involve surface principles in their encoding structure.

## SURFACE REPRESENTATION AS A RIEMANNIAN FIBER BUNDLE

Jun Zhang has made the interesting proposal that visual perception can be viewed as an interpretation based on the intrinsic geometry determined by rules of organization of the sensory data (Zhang and Wu, 1990; Zhang, 2005). The general idea is to relate perceptual unity to the concept of intrinsic constancy under a non-Euclidean geometry, which may be extended to visual modalities such as form, motion, color, and depth. The perceptual structure of the visual process can then be described as a fiber bundle, with visual space as the base manifold, the mapping from the world to the cortex as the base connection, the motion system as the tangent fiber, and all other relevant visual modalities as general fibers within the fiber bundle. The cross section of the fiber bundle is the information from the visual scene, an intrinsically invariant (parallel) portion of which represents a visual object. This concept can account for the unity of perceptual binding of the variety of different perceptual cues that are segregated early in the visual process.

## MULTIPLE SURFACE CUES

Studies of surface properties typically focus on surfaces represented by purely stereoscopic cues, but physical surfaces are almost always defined by multiple visual cues. Thus, it is important to treat the multimodal representation of surfaces as a perceptual primary, integrating the properties of the reflectance structures in the world into a unified surface representation:

$$\Phi(x,y) = f(d_S(x,y), d_D(x,y), d_M(x,y), d_T(x,y)...) \tag{0.9}$$

where $d_X(x,y)$ are the independent egocentric distances computed from each of the independent distance cues ($S$, luminance shading; $D$, binocular disparity; $M$, motion parallax; $T$, texture gradient, etc.), and $f()$ is the operative cue combination rule.

This expression says that the information from these diverse modalities is combined into a unitary depth percept, but it does not specify the combination rule by which they are aggregated. For the commonly proposed Bayesian combination rule,

$$f(\ ) = \left\| \sigma_X^2 / d_X \right\|_2, \tag{0.10}$$

where $\sigma_X^2$ are the noise variances for each distance cue.

If surface reconstruction is performed separately in each visual modality (with independent noise sources), the surface distance estimates should combine according to their absolute signal–noise ratios. Signals from the various modalities ($S$, $D$, $M$, $T$, ..., $X$) would combine to improve the surface distance estimation; adding information about the object profile from a second surface identification modality could never degrade surface reconstruction accuracy.

More realistic, post-Bayesian versions of the combination rule have also been proposed (Tyler, 2004). The main rule by which the area is assigned to a particular border is the occlusion rule. If a disparity-defined structure encloses a certain region, that region is seen as occluding the region outside the border. The border is perceived as "owning" the region inside the enclosure, which is therefore assigned to the same depth as the border. The region outside the enclosure is therefore more distant and is "owned" by the next depth-defined border that is reached beyond the first. These processes must somehow be implemented in the neural representation of the surfaces that we perceive on viewing these images.

In terms of Equation (0.10), the computational issue that needs to be faced is that the cue combination operation works only if each distance cue exists at every point $(x,y)$. However, the only distance cue that is continuously represented across the visual field is that of luminance shading. All the other cues depend for the computation of distance on the existence of local contrast, which in general is sparsely represented across the field. Computationally, therefore, the process would need to incorporate the ability to operate when some subset of the cues, or all the cues, have sparse values in a subset $(x,y)$ of directions in space, as in Equations (0.5) and (0.6). Moreover, there needs to be some mechanism of integrated interpolation in the process of the sparse cue combination.

## SURFACE REPRESENTATION THROUGH THE ATTENTIONAL SHROUD

One corollary of this surface reconstruction approach is a postulate that the object array is represented strictly in terms of its surfaces, as proposed by Nakayama and Shimojo (1990). Numerous studies point to a key role of surfaces in organizing the perceptual inputs into a coherent representation. Norman and Todd (1998), for example, show that depth discrimination is greatly improved if the two locations to be discriminated lie in a surface rather than being presented in empty space. This result is suggestive of a surface level of interpretation, although it may simply be relying on the fact that the presence of the surface provides more information about the depth regions to be assessed. Nakayama, Shimojo, and Silverman (1989) provide many demonstrations of the importance of surfaces in perceptual organization. Recognition of objects (such as faces) is much enhanced where the scene interpretation allows them to form parts of a continuous surface rather than isolated pieces, even when the retinal information about the objects is identical in the two cases.

A more vivid representation of the reconstruction process is to envisage it as an attentional shroud (Tyler and Kontsevich, 1995; Tyler, 2006), wrapping the dense locus of activated disparity detectors as a cloth wraps a structured object (see Figure 0.6). The concept of the attentional shroud is intended to capture the idea of a mechanism that acts like the soap film of Equation (0.7) in minimizing the curvature of the perceived depth surface consistent with

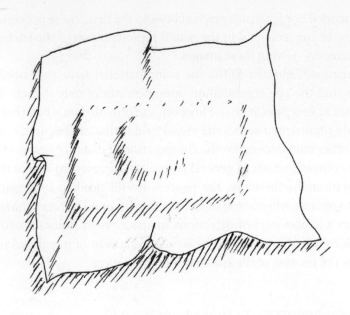

FIGURE 0.6 Depiction of the idea of an attentional shroud wrapping an object, here a camera. The information in the configuration of the shroud conveys the concept of the object shape in a coarse surface representation. The attentional shroud is conceived as a self-organizing manifold drawn to features of the object shape defined by depth cue representations somewhere in the cortex.

the available disparity information. Concepts of "mirror neurons" imply that there are neural representations of the actions of others implemented in the brain in a form that is both manipulable and adaptive to new situations faced by the observer (Rizzolati et al., 1996; Rizzolatti and Sinigaglia, 2010). The concept of the attentional shroud shares some similarities with the mirror concept, in that it is mirroring the configuration of the surface in the world with an internal surface representation that can match the configurational properties of the surfaces being viewed.

## EMPIRICAL EVIDENCE FOR SURFACE REPRESENTATION IN THE BRAIN

Surface representations are often discussed in terms of brightness propagation and texture segmentation, but these are weak inferences toward a true surface representation. Evidence is building from perceptual, psychophysical, neurophysiological, and computational sources in support of a surface-level description operating in the brain. Surface-specific neural coding has been reported early in the visual processing stream (Nienborg et al., 2005; Bredfeldt and Cumming, 2006; Samonds, Potetz, and Lee, 2009) and subsequently appears to be a feature of both the temporal-lobe and parietal-lobe streams of spatial representation in the cortex. In the temporal-lobe stream, neurons responsive to stereoscopic surface orientation have been reported in visual area V4 (Hinkle and Connor, 2002), and in middle temporal areas (MT) (Nguyenkim and DeAngelis, 2003) and medical superior temporal (MST) (Sugihara et al., 2002). Deeper into the temporal lobe, many neurons in the inferior bank of the superior temporal sulcus are selective for the complex shape of stereoscopic surfaces (Sakata et al., 1999; Janssen et al., 2001; Tanaka et al., 2001; Liu, Vogels, and Orban, 2004). Moreover, Joly, Vanduffel, and Orban. (2009) observed depth structure sensitivity from disparity in a small region of macaque inferior temporal cortex, TEs, known to house higher-order disparity selective neurons. Even in the frontal lobe, within ventral premotor cortex, area F5a (the most rostral sector of F5), showed sensitivity for depth structure from disparity. Within this area, 2D shape sensitivity was also observed, suggesting that area F5a processes complete 3D shape and might thus reflect the activity of canonical neurons.

Similarly, several regions of the parietal cortex are involved in the coding of surface shape. At a simple level, a large proportion of neurons in the occipital extension of the intraparietal sulcus of monkey are selective for stereoscopic surface orientation in the third dimension (Shikata et al., 1996; Taira et al., 2000; Tsutsui et al., 2001, 2002). This wealth of studies makes it clear that multimodal surface representation is an important component of the neural hierarchy in both the ventral and dorsal processing streams. Furthermore, Durand et al. (2007) used functional magnetic resonance imaging (fMRI) in monkeys to show that while several intraparietal (IP) areas (caudal, lateral, and anterior IP areas CIP, LIP, and AIP on the lateral bank; posterior and medial IP areas PIP and MIP on the medial bank) are activated by stereoscopic stimuli, AIP and an adjoining portion of LIP are sensitive to the stereoscopic surface shape of small objects. Interestingly, unlike the known representation of 3D shape in macaque inferior temporal cortex, the neural representation in AIP appears to emphasize object parameters required for the planning of grasping movements (Srivastava et al., 2009). This interpretation provides the basis for an understanding of the dual coding of surface shape in both the ventral and dorsal streams.

The dorsal stream would be involved in the 3D properties for the preparation for action, and the ventral stream would be specialized for the processes of semantic encoding and categorization of the objects.

In human cortex, Tyler et al. (2006) focused on the occipital cortex and found a dorsal region of the lateral occipital complex (LOC) to be specialized for the surface structure represented by both stereoscopic and kinetic depth cues. They argued that this region was the first level for the encoding of the generic surface structure of visual objects. Extending along the intraparietal sulcus (IPS), Durand et al. (2009) determined that retinotopic area V7 had a mixed sensitivity to both position in depth and generic depth structure, and the dorsal medial (DIPSM) and the dorsal anterior (DIPSA) regions of the IPS were sensitive to depth structure and not to position in depth. All three regions were also sensitive to 2D shape, indicating that they carry full 3D shape information. Similarly, Georgieva et al. (2009) report the involvement of five IPS regions as well as the dorsal LOC, the posterior inferior temporal gyrus (ITG), and ventral premotor cortex in the extraction and processing of a 3D shape from the depth surface structure of objects in the world.

## CONCLUSION

This overview emphasizes the view of spatial vision as an active process of object representation, in which a self-organization net of neural representation can reach through the array of local depth cues to form an integrated surface representation of the object structure in the physical world being viewed. Such a description is compatible with a realization in the neural networks of the parieto-occipital cortex rather than just an abstract cognitive schema. This conceptualization identifies an active neural coding process that goes far beyond the atomistic concept of local contour or disparity detectors across the field and that can account for some of the dynamic processes of our visual experience of the surface structure of the scene before us. Once the 3D surface structure is encoded, the nature of the elements in the scene can be segmented into the function units that we know as "objects."

The contributions to this book develop advanced theoretical and empirical approaches to all levels of the surface representation problem, from both the computational and neural implementation perspectives. These cutting-edge contributions run the gamut from the basic issue of the ground plane for surface estimation through midlevel analyses of the processes of surface segmentation to complex Riemannian space methods of representing and evaluating surfaces. Taken together, they represent a radical new approach to the thorny problem of determining the structure and interrelationships of the objects in the visual scene, and one that holds the promise of a decisive advance both in our capability of parsing scene information computationally and also in our understanding the coding of such information in the neuronal circuitry of the brain.

## ACKNOWLEDGMENT

The idea for this book integrating cognitive science and computational vision approaches to 3D surface reconstruction grew out of a workshop developed by Christopher Tyler and Jun Zhang and supported by the US Air Force Office of Scientific Research. Authors contributing to Ch 0, 1, 3, 7, 9 (Christopher Tyler, Tai-Sing Lee, David Gu, Alessandro Sarti, and Lora Likova) were partially supported by AFOSR grant #FA9550-09-0678 to Christopher Tyler.

# Contributors

**Yiannis Aloimonos**
Department of Computer Sciences
University of Maryland
College Park, Maryland

**Giovanna Citti**
Dipartimento di Matematica
Universita di Bologna
Bologna, Italy

**James Coughlan**
The Smith-Kettlewell Eye
    Research Institute
San Francisco, California

**Patrick Garrigan**
Department of Psychology
St. Joseph's University
Philadelphia, Pennsylvania

**David Xianfeng Gu**
Computer Science Department
State University of New York at
    Stony Brook
Stony Brook, New York

**Volodymyr V. Ivanchenko**
The Smith-Kettlewell Eye
    Research Institute
San Francisco, California

**Philip J. Kellman**
Department of Psychology
University of California, Los Angeles
Los Angeles, California

**Tai Sing Lee**
Computer Science Department
Center for the Neural Basis of
    Cognition
Carnegie Mellon University
Pittsburgh, Pennsylvania

**Yunfeng Li**
Department of Psychological
    Sciences
Purdue University
West Lafayette, Indiana

**Lora T. Likova**
Smith-Kettlewell Brain Imaging
    Center
The Smith-Kettlewell Eye
    Research Institute
San Francisco, California

**Feng Luo**
Department of Mathematics
Rutgers University
New Brunswick, New Jersey

**Ajay Mishra**
Department of Computer
  Sciences
University of Maryland
College Park, Maryland

**Evan Palmer**
Department of Psychology
Wichita State University
Wichita, Kansas

**Zygmunt Pizlo**
Department of Psychological
  Sciences
Purdue University
West Lafayette, Indiana

**Brian Potetz**
Department of Electrical
  Engineering and Computer
  Science
University of Kansas
Lawrence, Kansas

**Alessandro Sarti**
Centre de Recherche en
  Épistémologie Appliquēe
  (CREA)
Ecole Polytechenique
Paris, France

**Tadamasa Sawada**
Department of Psychological
  Sciences
Purdue University
West Lafayette, Indiana

**James T. Todd**
Department of Psychology
Ohio State University
Columbus, Ohio

**Christopher W. Tyler**
Smith-Kettlewell Brain Imaging
  Center
The Smith-Kettlewell Eye
  Research Institute
San Francisco, California

**Rüdiger von der Heydt**
Krieger Mind/Brain Institute
Johns Hopkins University
Baltimore, Maryland

**Shing-Tung Yau**
Mathematics Department
Harvard University
Cambridge, Massachusetts

**Wei Zeng**
Computer Science Department
State University of New York at
  Stony Brook
Stony Brook, New York

# CHAPTER 1

# Scene Statistics and 3D Surface Perception

Brian Potetz

Tai Sing Lee

## CONTENTS

## 1.1 INTRODUCTION

The inference of depth information from single images is typically performed by devising models of image formation based on the physics of light interaction and then inverting these models to solve for depth. Once inverted, these models are highly underconstrained, requiring many assumptions, such as Lambertian surface reflectance, smoothness of surfaces, uniform albedo, or lack of cast shadows. Little is known about the relative merits of these assumptions in real scenes. A statistical understanding of the joint distribution of real images and their underlying three-dimensional (3D) structure would allow us to replace these assumptions and simplifications with probabilistic priors based on real scenes. Furthermore, statistical studies may uncover entirely new sources of information that are not obvious from physical models. Real scenes are affected by many regularities in the environment, such as the natural geometry of objects, the arrangements of objects in space, natural distributions of light, and regularities in the position of the observer. Few current computer vision algorithms for 3D shape inference make use of these trends. Despite the potential usefulness of statistical models and the growing success of statistical methods in vision, few studies have been made into the statistical relationship between images and range (depth) images. Those studies that have examined this relationship in nature have uncovered meaningful and exploitable statistical trends in real scenes which may be useful for designing new algorithms in surface inference, and also for understanding how humans perceive depth in real scenes (Howe and Purves, 2002; Potetz and Lee, 2003; Torralba and Oliva, 2002). In this chapter, we highlight some results we obtained in our study on the statistical relationships between 3D scene structures and two-dimensional (2D) images and discuss their implications on understanding human 3D surface perception and its underlying computational principles.

## 1.2 CORRELATION BETWEEN BRIGHTNESS AND DEPTH

To understand the statistical regularities in natural scenes that allow us to infer 3D structures from their 2D images, we carried out a study to investigate the correlational structures between depth and light in natural scenes. We collected a database of coregistered intensity and high-resolution range images (corresponding pixels of the two images correspond to the same point in space) of over 100 urban and rural scenes. Scans were collected using the Riegl LMS-Z360 laser range scanner. The Z360 collects coregistered range and color data using an integrated charge-coupled device (CCD) sensor and a time-of-flight laser scanner with a rotating mirror. The scanner has a maximum range of 200 m and a depth accuracy of 12 mm. However, for each scene in our database, multiple scans were averaged to obtain accuracy under 6 mm. Raw range measurements are given in meters. All scanning is performed in spherical coordinates. Scans were taken of a variety of rural and urban scenes. All images were taken outdoors, under sunny conditions, while the scanner was level with the ground. Typical spatial resolution was roughly 20 pixels per degree.

To begin to understand the statistical trends present between 3D shape and 2D appearance, we start our statistical investigation by studying simple linear correlations within 3D scenes. We analyzed corresponding intensity and range patches, computing the correlation between a specific pixel (in either image or range patch) with other pixels in the image patch or the range patch, obtained with the following equation:

$$\rho = cor[X,Y] = \frac{\text{cov}[X,Y]}{\sqrt{\text{var}[X]\text{var}[Y]}} \quad (1.1)$$

The patch size is 25 × 25 pixels, slightly more than 1 degree visual angle in each dimension, and in calculating the covariance, both of the image patch and the range patch, we subtracted their corresponding means across all patches.

One significant source of variance between images is the intensity of the light source illuminating the scene. Differences in lighting intensity result in changes to the contrast of each image patch, which is equivalent to applying a multiplicative constant. In order to compute statistics that are invariant to lighting intensity, previous studies of the statistics of natural images (without range data) focus on the logarithm of the light intensity values, rather than intensity (Field, 1994; van Hateren and van der Schaaf, 1998). Zero-sum linear filters will then be insensitive to changes in image contrast. Likewise, we take the logarithm of range data as well. As explained by Huang et al. (2000), a large object and a small object of the same shape will appear identical to the eye when the large object is positioned appropriately far away and the small object is close. However, the raw range measurements of the large, distant object will differ from those of the small object by a constant multiplicative factor. In the log range data, the two objects will differ by an additive constant. Therefore, a zero-sum linear filter will respond identically to the two objects.

Figure 1.1 shows three illustrative correlation plots. Figure 1.1a shows the correlation between intensity at center pixel (13,13) and all of the pixels of the intensity patch. Figure 1.1b shows the correlation between range at pixel (13,13) and the pixels of the range patch. We observe that neighboring range pixels are much more highly correlated with one another than neighboring luminance pixels. This suggests that the low-frequency components of range data contain much more power than in luminance images, and that the spatial Fourier spectra for range images drop off more quickly than for luminance images, which are known to have roughly $\frac{1}{f}$ spatial Fourier amplitude spectra (Ruderman and Bialek, 1994). This finding is reasonable, because factors that cause high-frequency variation in range images, such as occlusion contours or surface texture, tend to also cause variation in the luminance image. However,

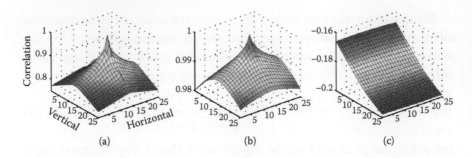

FIGURE 1.1 (a) Correlation between intensity at central pixel (13,13) and all of the pixels of the intensity patch. Note that pixel (1,1) is regarded as the upper-left corner of the patch. (b) Correlation between range at pixel (13,13) and the pixels of the range patch. (c) Correlation between intensity at pixel (13,13) and the pixels of the range patch. For example, correlation between intensity at central pixel (13,13) and lower-right pixel (25,25) was −0.210.

much of the high-frequency variations found in luminance images, such as shadow and surface markings, are not observed in range images. These correlations are related to the relative degree of smoothness characteristic of natural images versus natural range images. Specifically, natural range images are in a sense smoother than natural images. Accurate modeling of these statistical properties of natural images and range images is essential for robust computer vision algorithms and for perceptual inference in general. Smoothness properties in particular are ubiquitous in modern computer vision techniques for applications such as image denoising and inpainting (Roth and Black, 2005), image-based rendering (Woodford et al., 2006), shape from stereo (Scharstein and Szeliski, 2002), shape from shading (Potetz, 2007), and others.

Figure 1.1c shows correlation between intensity at pixel (13,13) and the pixels of the range patch. There are two important effects here. The first is a general vertical tilt in the correlation plot, showing that luminance values are more negatively correlated with depth at pixels lower within the patch. This result is due to the fact that the scenes in our database were lit from above. Because of this, surfaces facing upward were generally brighter than surfaces facing downward; conversely, brighter surfaces were more likely to be facing upward than darker surfaces. Thus, when a given pixel is bright, the distance to that pixel is generally less than the distance to pixels slightly lower within the image. This explains the increasingly negative correlations between the intensity at pixel (13,13) and the depth at pixels lower within the range image patch.

What is more surprising in Figure 1.1c is that the correlation between depth and intensity is significantly negative. Specifically, the correlation between the intensity and the depth at a given pixel is roughly −0.20. In other words, brighter pixels tend to be closer to the observer. Historically, physics-based approaches to shape from shading have generally concluded that shading cues offer only relative depth information. Our findings show there is also an absolute depth cue available from image intensity data that could help to more accurately infer depth from 2D images.

This empirical finding regarding natural 3D scenes may be related to an analogous psychophysical observation that, all other things being equal, brighter stimuli are perceived as being closer to the observer. This psychophysical phenomenon has been observed as far back as Leonardo da Vinci, who stated, "among bodies equal in size and distance, that which shines the more brightly seems to the eye nearer" (MacCurdy, 1938). Hence, we referred to our empirical correlation as the *da Vinci correlation*. Artists sometimes make use of this cue to help create compelling illusions of depth (Sheffield et al., 2000; Wallschlaeger and Busic–Snyder, 1992).

In the last century, psychophysicists validated da Vinci's observations in rigorous, controlled experiments (Ashley, 1898; Carr, 1935; Coules, 1955; Farne, 1977; Langer and Bülthoff, 1999; Surdick et al., 1997; Taylor and Sumner, 1945; Tyler, 1998; Wright and Ledgeway, 2004). In psychology literature, this effect is known as *relative brightness* (Myers 1995). Numerous possible explanations have been offered as to why such a perceptual bias exists. One common explanation is that light coming from distant objects has a greater tendency to be absorbed by the atmosphere (Cutting and Vishton, 1995). However, in most conditions, as in outdoor sunlit scenes, the atmosphere tends to scatter light from the sun directly toward our eyes, making more distant objects appear brighter under hazy conditions (Nayar and Narasimhan, 1999). Furthermore, our database was acquired under sunny, clear conditions, under distances insufficient to cause atmospheric effects (maximum distances were roughly 200 m). Other explanations of a purely psychological explanation have also been advanced (Taylor and Sumner, 1945). Although these might be contributing factors for our perceptual bias, they cannot account for empirical observations of real scenes.

By examining which images exhibited the da Vinci correlation most strongly, we concluded that the major cause of the correlation was primarily due to shadow effects within the environment (Potetz and Lee, 2003). For example, one category of images where correlation between nearness and brightness was very strong was images of trees and leafy foliage. Because the

FIGURE 1.2 **(See color insert.)** An example color image (top) and range image (bottom) from our database. For purposes of illustration, the range image is shown by displaying depth as shades of gray. Notice that dark regions in the color image tend to lie in shadow, and that shadowed regions are more likely to lie slightly farther from the observer than the brightly lit outer surfaces of the rock pile. This example image from our database had an especially strong correlation between closeness and brightness.

source of illumination comes from above, and outside of any tree, the outermost leaves of a tree or bush are typically the most illuminated. Deeper into the tree, the foliage is more likely to be shadowed by neighboring leaves, and so nearer pixels tend to be brighter. This same effect can cause a correlation between nearness and brightness in any scene with complex surface concavities and interiors. Because the light source is typically positioned outside of these concavities, the interiors of these concavities tend to be in shadow and more dimly lit than the object's exterior. At the same time, these concavities will be farther away from the viewer than the object's exterior. Piles of objects (such as Figure 1.2) and folds in clothing and fabric are other good examples of this phenomenon.

To test our hypothesis, we divided the database into urban scenes (such as building facades and statues) and rural scenes (trees and rocky terrain).

The urban scenes contained primarily smooth, man-made surfaces with fewer concavities or crevices, and so we predicted these images to have reduced correlation between nearness and brightness. On the other hand, were the correlation found in the original dataset due to atmospheric effects, we would expect the correlation to exist equally well in both the rural and urban scenes. The average depth in the urban database (32 m) was similar to that of the rural database (40 m), so atmospheric effects should be similar in both datasets. We found that correlations calculated for the rural dataset increased to −0.32, while those for the urban dataset are considerably weaker, in the neighborhood of −0.06.

In Langer and Zucker (1994)], it was observed that for continuous Lambertian surfaces of constant albedo, lit by a hemisphere of diffuse lighting and viewed from above, a tendency for brighter pixels to be closer to the observer can be predicted from the equations for rendering the scene. Intuitively, the reason for this is that under diffuse lighting conditions, the brightest areas of a surface will be those that are the most exposed to the sky. When viewed from above, the peaks of the surface will be closer to the observer. Although these theoretical results have not been extended to more general environments, our results show that in natural scenes, these tendencies remain, even when scenes are viewed from the side, under bright light from a single direction, and even when that lighting direction is oblique to the viewer. In spite of these differences, both phenomena seem related to the observation that concave areas are more likely to be in shadow. The fact that all of our images were taken under cloudless, sunny conditions and with oblique lighting from above suggests that this cue may be more important than first realized.

It is interesting to note that the correlation between nearness and brightness in natural scenes depends on several complex properties of image formation. Complex 3D surfaces with crevices and concavities must be present, and cast shadows must be present to fill these concavities. Additionally, we expect that without diffuse lighting and lighting inter-reflections (light reflecting off of several surfaces before reaching the eye), the stark lighting of a single-point light source would greatly diminish the effect (Langer and Zucker, 1994). Cast shadows, complex 3D surfaces, diffuse lighting, and lighting interreflections are image formation phenomena that are traditionally ignored by methods of depth inference that attempt to invert physical models of image formation. The mathematics required for these phenomena are too cumbersome to invert. However, taken together, these image formation behaviors result in the simplest possible relationship between shape and appearance—an absolute correlation between nearness and brightness. This finding illustrates the necessity of continued exploration of the statistics of natural 3D scenes.

## 1.3 CHARACTERIZING THE LINEAR STATISTICS OF NATURAL 3D SCENES

In the previous section, we explained the correlation between the intensity of a pixel and its nearness. We now expand this analysis to include the correlation between the intensity of a pixel and the nearness of other pixels in the image. The set of all such correlations forms the cross-correlation between depth and intensity. The cross-correlation is an important statistical tool: as we explain later, if the cross-correlation between a particular image and its range image was known completely, then given the image, we could use simple linear regression techniques to infer 3D shape perfectly. Even though perfect estimation of the cross-correlation from a single image is impossible, we demonstrate that this correlational structure of a single scene follows several robust statistical trends. These trends allow us to approximate the full cross-correlation of a scene using only three parameters, and these parameters can be measured even from very sparse shape and intensity information. Approximating the cross-correlation this way allows us to achieve a novel form of statistically driven depth inference that can be used in conjunction with other depth cues, such as stereo.

Given an image $i(x, y)$ with range image $z(x, y)$, the cross-correlation for that particular scene is given by

$$(i \star z)(\Delta x, \Delta y) = \int \int i(x, y) z(x{+}\Delta x, y{+}\Delta y) dx \, dy \qquad (1.2)$$

It is helpful to consider the cross-correlation between intensity and depth within the Fourier domain. If we use $I(u, v)$ and $Z(u, v)$ to denote the Fourier transform of $i(x, y)$ and $z(x, y)$, respectively, then the Fourier transform of $i, z$ is $Z(u, v)I^*(u, v)$. $ZI^*$ is known as the *cross-spectrum* of $i$ and $z$. Note that $ZI^*$ has both real and imaginary parts. Also note that in this section, no logarithm or other transformation was applied to the intensity or range data (measured in meters). This allows us to evaluate $ZI^*$ in the context of the Lambertian model assumptions, as we demonstrate later.

If the cross-spectrum is known for a given image, and is sufficiently bounded away from zero, then the 3D shape could be estimated from a single image using linear regression: $Z = I(ZI/II^*)$. In this section, we

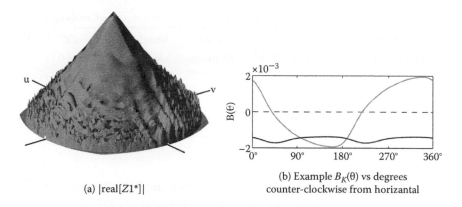

(a) |real[Z1*]|

(b) Example $B_K(\theta)$ vs degrees counter-clockwise from horizantal

FIGURE 1.3   (a) The log-log polar plot of | $real[ZI^*(r, \theta)]$ | for a scene from our database. (b) $B(\theta)$ for the same scene. $real[B_K(\theta)]$ is drawn in black and $imag[B_K(\theta)]$ in gray. This plot is typical of most scenes in our database. As predicted by Equation 1.5, $imag[B_K(\theta)]$ reaches its minima at the illumination direction (in this case, to the extreme left, almost 180°). Also typical is that $real[B_K(\theta)]$ is uniformly negative, most likely caused by cast shadows in object concavities. (Potetz and Lee, 2006.)

demonstrate that given only three parameters, a close approximation to $ZI^*$ can be constructed. Roughly speaking, those three parameters are the strength of the nearness/brightness correlation in the scene, the prevalence of flat shaded surfaces in the scene, and the dominant direction of illumination in the scene. This model can be used to improve depth inference in a variety of situations.

Figure 1.3a shows a log-log polar plot of | $real[ZI^*(r, \theta)]$ | from one image in our database. The general shape of this cross-spectrum appears to closely follow a power law. Specifically, we found that $ZI^*$ can be reasonably modeled by $B(\theta)/r^a$, where $r$ is spatial frequency in polar coordinates, and $B(\theta)$ is a function that depends only on polar angle $\theta$, with one curve for the real part and one for the imaginary part. We test this claim by dividing the Fourier plane into four 45° octants (vertical, forward diagonal, horizontal, and backward diagonal) and measuring the drop-off rate in each octant separately. For each octant, we average over the octant's included orientations and fit the result to a power law. The resulting values of $\alpha$ (averaged over all 28 images) are listed in Table 1.1.

For each octant, the correlation coefficient between the power-law fit and the actual spectrum ranged from 0.91 to 0.99, demonstrating that each octant is well-fit by a power law. (Note that averaging over orientation

TABLE 1.1    Power Law Drop-Off Rates $\alpha$ for Each Power Spectrum Component

| Orientation | II | Real[ZI] | Imag[ZI] | ZZ |
|---|---|---|---|---|
| Horizontal | $2.47 \pm 0.10$ | $3.61 \pm 0.18$ | $3.84 \pm 0.19$ | $2.84 \pm 0.11$ |
| Forward diagonal | $2.61 \pm 0.11$ | $3.67 \pm 0.17$ | $3.95 \pm 0.17$ | $2.92 \pm 0.11$ |
| Vertical | $2.76 \pm 0.11$ | $3.62 \pm 0.15$ | $3.61 \pm 0.24$ | $2.89 \pm 0.11$ |
| Backward diagonal | $2.56 \pm 0.09$ | $3.69 \pm 0.17$ | $3.84 \pm 0.23$ | $2.86 \pm 0.10$ |
| Mean | $2.60 \pm 0.10$ | $3.65 \pm 0.14$ | $3.87 \pm 0.16$ | $2.88 \pm 0.10$ |

smoothes out some fine structures in each spectrum.) Furthermore, $\alpha$ varies little across orientations, showing that our model fits $ZI^*$ closely.

Note from the table that the image power spectra $I(u, v)I^*(u, v)$ also obeys a power law. The observation that the power spectrum of natural images obeys a power law is one of the most robust and important statistic trends of natural images (Ruderman and Bialek, 1994), and it stems from the scale invariance of natural images. Specifically, an image that has been scaled up, such as $i(\sigma x, \sigma y)$, has similar statistical properties as an unscaled image. This statistical property predicts that $II^* (r, \theta) \approx 1/r^2$. The power-law structure of the power spectrum $II^*$ has proven highly useful in image processing and computer vision and has led to advances in image compression, image denoising, and several other applications. Similarly, the discovery that $ZI^*$ also obeys a power spectrum may prove highly useful for the inference of 3D shape.

As mentioned earlier, knowing the full cross-covariance structure of an image/range-image pair would allow us to reconstruct the range image using linear regression via the equation $Z = I(ZI^*/II^*)$. Thus, we are especially interested in estimating the regression kernel $K = ZI^*/II^*$. $IK$ is a perfect reconstruction of the original range image (as long as $II^*(u, v) \neq 0$). The findings shown in Table 1.1 predict that $K$ also obeys a power law. Subtracting $\alpha_{II^*}$ from $\alpha_{\text{real}[ZI^*]}$ and $\alpha_{\text{imag}[ZI^*]}$, we find that $real[K]$ drops off at $1/r^{1.1}$, and $imag[K]$ drops off at $1/r^{1.2}$. Thus, we have that $K(r, \theta) \approx BK(\theta)/r$.

Now that we know that $K$ can be fit (roughly) by a $1/r$ power law, we can offer some insight into why $K$ tends to approximate this general form. Note that the $1/r$ drop-off of $K$ cannot be predicted by scale invariance. If images and range images were jointly scale invariant, then $II^*$ and $ZI^*$ would both obey $1/r^2$ power laws, and $K$ would have a roughly uniform magnitude. Thus, even though natural *images* appear to be statistically scale invariant, the finding that $K \approx B_K(\theta)/r$ disproves scale invariance for natural *scenes* (meaning images and range images taken together).

The $1/r$ drop-off in the *imaginary* part of $K$ can be explained by the linear Lambertian model of shading, with oblique lighting conditions. Recall that Lambertian shading predicts that pixel intensity is given by

$$i(x,y) \propto \vec{n}(x,y) \cdot \vec{L} \tag{1.3}$$

where $\vec{n}(x,y)$ is the unit surface normal at point $(x, y)$, and $\vec{L}$ is the unit lighting direction. The linear Lambertian model is obtained by taking only the linear terms of the Taylor series of the Lambertian reflectance equation. Under this model, if constant albedo and illumination conditions are assumed, and lighting is from above, then $i(x, y) = a\, \partial z/\partial y$, where $a$ is some constant. In the Fourier domain, $I(u, v) = a2\pi jvZ(u, v)$, where $j = \sqrt{-1}$. Thus, we have that

$$ZI^{\star}(r,\theta) = -\frac{j}{a2\pi r\sin(\theta)} II^{\star}(r,\theta) \tag{1.4}$$

$$K(r,\theta) = -j\frac{1}{r}\frac{1}{a2\pi\sin(\theta)} \tag{1.5}$$

Thus, Lambertian shading predicts that $imag[ZI^*]$ should obey a power law, with $\alpha_{imag[ZI^*]}$ being one more than $\alpha_{imag[II^*]}$, which is consistent with the findings in Table 1.1.

Equation (1.4) predicts that only the imaginary part of $ZI^*$ should obey a power law, and the real part of $ZI^*$ should be zero. Yet, in our database, the real part of $ZI^*$ was typically stronger than the imaginary part. The real part of $ZI^*$ is the Fourier transform of the even-symmetric part of the cross-correlation function, and it includes the direct correlation $cov[i,z]$, corresponding to the da Vinci correlation between intensity pixel and range pixel discussed earlier. The form of $real[ZI^*]$ is related to the rate at which the da Vinci correlation drops off over space. One explanation for the $1/r^3$ drop-off rate of $real[ZI^*]$ is the observation that deeper crevices and concavities should be more shadowed and therefore darker than shallow concavities. Specifically, the $1/r^3$ drop-off rate of $real[ZI^*]$ predicts that the correlation between nearness and brightness should be inversely proportional to the aperture width it is measured over. This matches our intuition that given two surface concavities with equal depths, the one with the narrower aperture should be darkest.

Figure 1.4 shows examples of $B_K$ in urban and rural scenes. These plots illustrate that the real part of $B_K$ is strongest (most negative) for rural scenes with abundant concavities and shadows. These figures also illustrate how

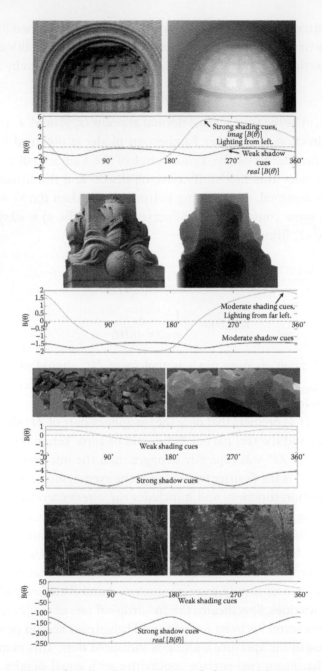

FIGURE 1.4 Natural and urban scenes and their $B_K(\theta)$. Images with surface concavities and cast shadows have significantly negative $real[B(\theta)]$ (black line), and images with prominent flat shaded surfaces have strong $imag[B(\theta)]$ (gray line).

the imaginary part of $K$ follows Equation (1.5), and imag$[B_K(\theta)]$ closely follows a sinusoid with phase determined by the dominant illumination angle. Thus, $B_K$ (and therefore also $K$ and $ZI^*$) can be approximated using only three parameters: the strength of the da Vinci correlation (which is related to the extent of complex 3D surfaces and shadowing present in the scene), the angle of the dominant lighting direction, and the strength of the Lambertian relationship in the scene (i.e., the coefficient $1/\alpha$ in Equation (1.5), which is related to the prominence of smooth Lambertian surfaces in the scene). We can use this approximation to improve depth inference.

## 1.4 IMPLICATIONS TOWARD DEPTH INFERENCE

Armed with a better understanding of the statistics of real scenes, we are better prepared to develop successful depth inference algorithms. One example is range image superresolution. Often, we may have a high-resolution color image of a scene but only a low spatial resolution range image (range images record the 3D distance between the scene and the camera for each pixel). This often happens if our range image was acquired by applying a stereo depth inference algorithm. Stereo algorithms rely on smoothness constraints, either explicitly or implicitly, and so the high-frequency components of the resulting range image are not reliable (Cryer et al., 1995; Scharstein and Szeliski, 2002). Laser range scanners are another common source of low-resolution range data. Laser range scanners typically acquire each pixel sequentially, taking up to several minutes for a high-resolution scan. These slow scan times can be impractical in real situations, so in many cases, only sparse range data are available. In other situations, inexpensive scanners are used that can capture only sparse depth values.

It should be possible to improve our estimate of the high spatial frequencies of the range image by using monocular cues from the high-resolution intensity (or color) image. One recent study suggested an approach to this problem known as *shape recipes* (Freeman and Torralba, 2003; Torralba and Freeman, 2003). The basic principle of shape recipes is that a relationship between shape and appearance could be *learned* from the low-resolution image pair, and then *extrapolated* and applied to the high-resolution intensity image to infer the high spatial frequencies of the range image. One advantage of this approach is that hidden variables important to inference from monocular cues, such as illumination direction and material reflectance properties, might be implicitly learned from the low-resolution range and intensity images.

From our statistical study, we now know that fine details in $K = ZI^*/II^*$ do not generalize across scales, as was assumed by shape recipes. However, the coarse structure of $K$ roughly follows a $1/r$ power law. We can exploit this statistical trend directly. We can simply estimate $B_K(\theta)$ using the low-resolution range image, use the $1/r$ power law to extrapolate $K \approx B_K(\theta)/r$ into the higher spatial frequencies, and then use this estimate of $K$ to reconstruct the high-frequency range data. Specifically, from the low-resolution range and intensity image, we compute low-resolution spectra of $ZI^*$ and $II^*$. From the highest frequency octave of the low-resolution images, we estimate $B_{II}(\theta)$ and $B_{ZI}(\theta)$. Any standard interpolation method will work to estimate these functions. We chose a $cos^3(\theta + \pi\phi/4)$ basis function based on steerable filters (Adelson and Freeman, 1991). We now can estimate the high spatial frequencies of the range image, $z$. Define

$$K_{powerlaw}(r,\theta) = (B_{ZI}(\theta)/B_{II}(\theta))/r \qquad (1.6)$$

$$Z_{powerlaw} = F_{low}(r)Z + (1 - F_{low}(r))IK_{powerlaw} \qquad (1.7)$$

where $F_{low}$ is the low-pass filter that filters out the high spatial frequencies of $z$ where depth information is either unreliable or missing.

Because our model is derived from scene statistics and avoids some of the mistaken assumptions in the original shape recipe model, our extension provides a twofold improvement over Freeman and Torralba's original approach (2003), while using far fewer parameters. Figure 1.5 shows an example of the output of the algorithm.

This power law–based approach can be viewed as a statistically informed generalization of a popular shape-from-shading algorithm known as *linear shape from shading* (Pentland, 1990), which remains popular due to its high efficiency. Linear shape from shading attempts to reconstruct 3D shape from a single image using Equation (1.5) alone, ignoring shadow cues and the da Vinci correlation. As mentioned previously, the da Vinci correlation is a product of cast shadows, complex 3D surfaces, diffuse lighting, and lighting interreflections. All four of these image formation phenomena are exceptionally cumbersome to invert in a deterministic image formation model, and subsequently, they have been ignored by most previous depth inference algorithms. However, taken together, these phenomena produce a simple statistical relationship that can be exploited using highly efficient linear algorithms such as Equation 1.7. It was not until the statistics of natural range and intensity images were studied empirically that the strength of these statistical cues was made clear.

(a) Original intensity image    (b) Low-resolution range data    (c) Power-law technique

FIGURE 1.5 (a) An example intensity image from our database. (b) A computer-generated Lambertian rendering of the corresponding laser-acquired low-resolution range image. This figure shows the low-resolution range image that, for purposes of illustration, has been artificially rendered as an image. Note the oversmoothed edges and lack of fine spatial details that result from the downsampling. (c) Power-law method of inferring high-resolution three-dimensional (3D) shape from a low-resolution range image and a high-resolution color image. High spatial-frequency details of the 3D shape have been inferred from the intensity image (left). Notice that some high-resolution details, such as the cross in the sail, are not present in the low-resolution range image but were inferred from the full-resolution intensity image. (Potetz and Lee, 2006.)

The power-law algorithm described here presents a new opportunity to test the usefulness of the da Vinci shadow cues, by comparing the power-law algorithm results to the linear shape-from-shading technique (Pentland, 1990). When our algorithm was made to use only shading cues (by setting the real part of $K_{powerlaw}(r, \theta)$ to zero), the effectiveness of the algorithm was reduced to 27% of its original performance. When only shadow cues were used (by setting the imaginary part of $K_{powerlaw}(r, \theta)$ to zero), the algorithm retained 72% of its original effectiveness (Potetz and Lee, 2006). Thus, in natural scenes, linear shadow cues proved to be significantly more powerful than linear shading cues. These results show that shadow cues are far more useful than was previously expected. This is an important empirical observation, as shape from shading has received vastly more attention in computer vision research than shape from shadow. This finding highlights the importance of shadow cues and the benefits of statistical studies of natural scenes.

As expected given the analysis of the da Vinci correlation above, the relative performance of shadow and shading cues depends strongly on the category of the images considered. Shadow cues were responsible for 96% of algorithm performance in foliage scenes, 76% in scenes of rocky terrain, and 35% in urban scenes.

## 1.5 STATISTICAL INFERENCE FOR DEPTH INFERENCE

The approach described above shows the limits of what is possible using only second-order linear statistics. The study of these simple models is important, because it helps us to understand the statistical relationships that exist between shape and appearance. However, simple linear systems capture only a fraction of what is achievable using a complete statistical inference framework. The problem of inferring a 3D shape from image cues is both highly complex and highly underconstrained: for any given 2D image, there are countless plausible 3D interpretations of that scene. Our goal is to find solutions that are especially likely. Powerful statistical methods will be necessary to achieve these goals. In this section, we discuss the use of modern statistical inference techniques for inferring a 3D shape from images.

In recent years, there has been a great deal of progress made in computer vision using graphical models of large joint probability distributions (Fei-Fei and Perona, 2005; Freeman et al., 2000; Sun et al., 2003; Tang et al., 2005; Torralba et al., 2003; Zhu and Tu, 2002). Graphical models offer a powerful framework to incorporate rich statistical cues from natural scenes and can be applied directly to the problem of depth inference. Bayesian inference of shape (depth) $Z$ from images $I$ involves estimating properties of the posterior distribution $P(Z \mid I)$. The dimensionality of the posterior distribution $P(Z \mid I)$, however, is far too great to model directly. An important observation relevant to vision is that the interdependency of variables tends to be relatively local. This allows the factorization of the joint distribution into a product of "potential functions," each of lower dimensionality than the original distribution (as shown in Figure 1.6). In other words,

$$P(I,Z) \propto \prod_a \phi_a(\vec{x}_a) \qquad (1.8)$$

where $\vec{x}_a$ is some subset of variables in $I$ and $Z$. Such a factorization defines an example of a graphical model known as a "factor graph": a bipartite graph with a set of variable nodes (one for each random variable in the

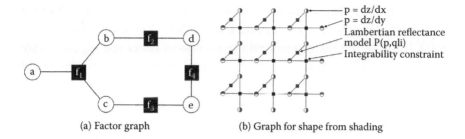

(a) Factor graph          (b) Graph for shape from shading

FIGURE 1.6   (a) An example factor graph. This graph represents the factorization of a joint probability distribution over five random variables: $P(a, b, c, d, e) \propto f_1(a, b, c) f_2(b, d) f_3(c, e) f_4(d, e)$. (b) A factor graph to solve the classical Lambertian shape-from-shading problem using linear constraint nodes. The representation of a three-dimensional (3D) shape is twice over-complete, including $p$ and $q$ slope values at each pixel. The linear constraint nodes are shown as black squares, and they enforce the consistency (integrability) of the solution. The gray squares represent factor nodes encoding the reflectance function.

multivariate distribution) and a set of factor nodes (one for each potential function). Each factor node is connected to each variable referenced by its corresponding potential function. (See Figure 1.6 for an example, or reference (Frey, 1998) for a review of factor graphs.) Factor graphs that satisfy certain constraints can be expressed as Bayes networks, or for other constraints, as Markov random fields (MRFs). Thus, these approaches are intimately connected and are equivalent in terms of neural plausibility.

Exact inference on factor graphs is possible only for a small subclass of problems. In most cases, approximate methods must be used. There are a variety of existing approaches to approximate the mode of the posterior distribution (MAP, or maximum *a posteriori*) or its mean (MMSE, or minimum mean-squared error), such as Markov chain Monte Carlo (MCMC) sampling, graph cuts, and belief propagation. In this section, we explore the use of the belief propagation algorithm. Belief propagation is advantageous in that it imposes fewer restrictions on the potential functions than graph cuts (Kolmogorov and Zabih, 2001) and is faster than MCMC. Belief propagation is also interesting in that it is highly neurally plausible (Lee and Mumford, 2003; Rao, 2004) and has been advanced as a possible model for statistical inference in the brain (Pearl, 1988).

Belief propagation has been applied successfully to a wide variety of computer vision problems (Freeman et al., 2000; Potetz, 2007; Sun et al.,

2003; Tang et al., 2005) and has shown impressive empirical results on a number of other problems (Frey and Dueck, 2007; Kschischang and Frey, 1998). Initially, the reasons behind the success of belief propagation were only understood for those cases where the underlying graphical model did not contain loops. The many empirical successes on graphical models that did contain loops were largely unexplained. However, recent discoveries have provided a solid theoretical justification for "loopy" belief propagation by showing that when belief propagation converges, it computes a minima of a measure used in statistical physics known as the Bethe free energy (Yedidia et al., 2000). The Bethe free energy is based on a principled approximation of the KL-divergence between a graphical model and a set of marginals and has been instrumental in studying the behaviors of large systems of interacting particles, such as spin glasses. The connection to Bethe free energy had an additional benefit in that it inspired the development of algorithms that minimize the Bethe free energy directly, resulting in variants of belief propagation that guarantee convergence (Heskes et al., 2003; Yuille, 2002), improve performance (Yedidia et al., 2000, 2003), or in some cases, guarantee that belief propagation computes the *globally* optimal MAP point of a distribution (Weiss and Freeman, 2007).

Belief propagation estimates the marginals $b_i(x_i) = \Sigma_{X \setminus x_i} P(\vec{X})$ by iteratively computing *messages* a long each edge of the graph according to the following equations:

$$m_{i \to f}^{t+1}(x_i) = \prod_{g \in \mathcal{N}(i) \setminus f} m_{g \to i}^t(x_i) \tag{1.9}$$

$$m_{f \to i}^{t+1}(x_i) = \sum_{\vec{x}_{\mathcal{N}(f) \setminus i}} \left( \phi_f\left(\vec{x}_{\mathcal{N}(f)}\right) \prod_{j \in \mathcal{N}(f) \setminus i} m_{j \to f}^t(x_j) \right) \tag{1.10}$$

$$b_i(x_i) \propto \prod_{g \in \mathcal{N}(i)} m_{g \to i}^t(x_i) \tag{1.11}$$

where $f$ and $g$ are factor nodes, $i$ and $j$ are variable nodes, and $\mathcal{N}(i)$ is the set of neighbors of node $i$ (Heskes, 2004). Here, $bi(xi)$ is the estimated marginal of variable $i$. Note that the expected value of $\vec{X}$ or equivalently, the minimum mean-squared error (MMSE) point estimate, can be computed by finding the mean of each marginal. If the most likely value of $\vec{X}$ is desired, also known as the maximum *a posteriori* (MAP) point estimate, then the integrals of Equation (1.10) are replaced by suprema. This is known as *max-product belief propagation*.

For many computer vision problems, belief propagation is prohibitively slow. The computation of Equation (1.10) has a complexity of $\mathcal{O}(M^N)$, where $M$ is the number of possible labels for each variable, and $N$ is the number of neighbors of factor node $f$. In many computer vision problems, variables are continuous or have many labels. In these cases, applications of belief propagation have nearly always been restricted to pairwise connected MRFs, where each potential function depends on only two variable nodes (i.e., $N = 2$) (Freeman et al., 2000; Sun et al., 2003). However, pairwise connected models are often insufficient to capture the full complexity of the joint distribution and thus would severely limit the expressive power of factor graphs. Developing efficient methods for computing nonpairwise belief propagation messages over continuous random variables is therefore crucial for solving complex problems with rich, higher-order statistical distributions encountered in computer vision.

In the case that the potential function $\phi$ can be expressed in terms of a weighted sum of its inputs, we developed a set of techniques to speed up the computation of messages considerably. For example, suppose the random variables $a$, $b$, $c$, and $d$ are all variable nodes in our factor graph, and we want to constrain them such that $a + b = c + d$. We would add a factor node $f$ connected to all four variables with potential function

$$\phi_f(a,b,c,d) = \delta(a+b-c-d) \tag{1.12}$$

To compute $m_{f \to A}^{t+1}$, we use Equation 1.10:

$$m_{f \to A}^{t+1}(a) = \sum_{b,c,d} \delta(a+b-c-d) m_{B \to f}^t(b) m_{C \to f}^t(c) m_{D \to f}^t(d) \tag{1.13}$$

$$= \sum_{b,c} m_{B \to f}^t(b) m_{C \to f}^t(c) m_{D \to f}^t(a+b-c) \tag{1.14}$$

$$= \sum_{x,y} m_{B \to f}^t(x-a) m_{C \to f}^t(x-y) m_{D \to f}^t(y) \tag{1.15}$$

$$= \sum_{x} m_{B \to f}^t(x-a) \left( \sum_{y} m_{C \to f}^t(x-y) m_{D \to f}^t(y) \right) \tag{1.16}$$

where $x = a + b$ and $y = a + b - c$. Notice that in Equation (1.16), the second summand (in parentheses) does not depend on $a$. This summand can be computed in advance by summing over $y$ for each value of $x$. Thus, computing $m_{f \to A}^{t+1}(a)$ using Equation (1.16) is $\mathcal{O}(M^2)$, which is far superior to a straightforward computation of Equation (1.13), which is $\mathcal{O}(M^4)$. In (Potetz and Lee, 2008), we show how this approach can be used to compute messages in time $\mathcal{O}\ (M^2)$ for all potential functions of the form

$$\phi(\vec{x}) = g\left(\sum_i g_i(x_i)\right) \qquad (1.17)$$

This reduces a problem from exponential time to linear time with respect to the number of variables connected to a factor node. Potentials of this form are common in graphical models, in part because they offer advantages in training graphical models from real data (Friedman et al., 1984; Hinton, 1999; Roth and Black, 2005; Zhu et al., 1998).

This approach reduces a problem from exponential time to linear time with respect to the number of variables connected to a factor node. With this efficient algorithm, we were able to apply belief propagation to the classical computer vision problem of shape from shading, using the factor graph shown in Figure 1.6 (see Potetz, 2007 for details). Previously, the general problem of shape from shading was solved using gradient descent–based techniques. In complex, highly nonlinear problems like shape from shading, these approaches often become stuck inside local, suboptimal minima. Belief propagation helps to avoid difficulties with local minima in part because it operates over whole probability distributions. Gradient descent approaches maintain only a single 3D shape at a time, iteratively refining that shape over time. Belief propagation seeks to optimize the single-variate marginals $b_i(x_i)$ for each variable in the factor graph.

Solving shape from shading using belief propagation performs significantly better than previous state-of-the-art techniques (see Figure 1.7). Note that without the efficient techniques described here, belief propagation would be intractable for this problem, requiring over 100,000 times longer to compute each iteration. In addition to improved performance, solving shape from shading using belief propagation allows us to relax many of the restrictions typically assumed by shape-from-shading algorithms in order to make the problem tractable. The classical definition of the shape-from-shading problem specifies that lighting must originate from a single point source, that surfaces should be entirely matte, or Lambertian in reflectance, and that no markings or colorations can be present on any surface. The flexibility of the belief propagation approach allows us to start relaxing these constraints, making shape from shading viable in more realistic scenarios.

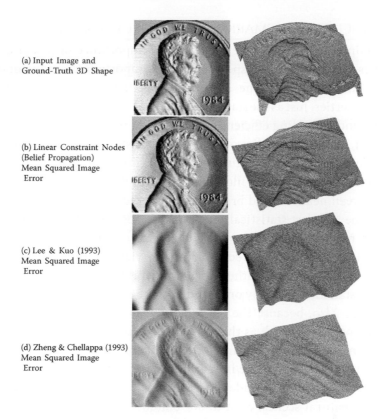

(a) Input Image and Ground-Truth 3D Shape

(b) Linear Constraint Nodes (Belief Propagation) Mean Squared Image Error

(c) Lee & Kuo (1993) Mean Squared Image Error

(d) Zheng & Chellappa (1993) Mean Squared Image Error

FIGURE 1.7  Comparison between our results of inferring shape from shading using loopy belief propagation (row b) with previous approaches (rows c and d). Each row contains a three-dimensional (3D) wire mesh plot of the surface (right) and a rendering (left) of that surface under a light source at location (1,0,1). (a) The original surface. The rendering in this column serves as the input to the shape from shading (SFS) algorithms in the next three columns. (b) The surface recovered using our linear constraint node approach. (c) The surface recovered using the method described by Lee and Kuo (1993). This algorithm performed best of the six SFS algorithms reviewed in a survey paper by Zhang et al. (1999). (d) The surface recovered using the method described by Zheng and Chellappa (1991). Our approach (row b) offers a significant improvement over previous leading methods. It is especially important that rerendering that recovered surface very closely resembles the original input image. This means that the Lambertian constraint at each pixel was satisfied, and that any error between the original and recovered surface is primarily the fault of the simplistic model of prior probability of natural 3D shapes used here.

## 1.6 CONCLUDING REMARKS AND FUTURE DIRECTIONS

The findings described here emphasize the importance of studying the statistics of natural scenes; specifically, it is important to study not only the statistics of images alone, but images together with their underlying scene properties. Just as the statistics of natural images has proven invaluable for understanding efficient image coding, transmission, and representation, the joint statistics of natural scenes stands to greatly advance our understanding of perceptual inference. The discovery of the da Vinci correlation described here illustrates this point. This absolute correlation between nearness and brightness observed in natural 3D scenes is among the simplest statistical relationships possible. However, it stems from the most complex phenomena of image formation; phenomena that have historically been ignored by computer vision approaches to depth inference for the sake of mathematical tractability. It is difficult to anticipate this statistical trend by only studying the physics of light and image formation. Also, because the da Vinci correlation depends on intrinsic scene properties such as the roughness or complexity of a 3D scene, physical models of image formation are unable to estimate the strength of this cue, or its prevalence in real scenes. By taking explicit measurements using laser range finders, we demonstrated that this cue is very strong in natural scenes, even under oblique, nondiffuse lighting conditions. Further, we showed that for linear depth inference algorithms, shadow cues such as the da Vinci correlation are 2.7 times as informative as shading cues in a diverse collection of natural scenes. This result is especially significant, because depth cues from shading have received far more attention than shadow cues. We believe that continued investigation into natural scene statistics will continue to uncover important new insights into visual perception that are unavailable to approaches based on physical models alone.

Another conclusion we wish to draw is the benefit of statistical methods of inference for visual perception. The problem of shape from shading described above was first studied in the 1920s in order to reconstruct the 3D shapes of lunar terrains (Horn, 1989). Since that time, approaches to shape from shading were primarily deterministic and typically involved iteratively refining a single shape hypothesis until convergence was reached. By developing and applying efficient statistical inference techniques that consider *distributions* over 3D shapes, we were able to advance the state of shape from shading considerably.

The efficient belief propagation techniques we developed have similar applications in a variety of perceptual inference tasks. These and other statistical inference techniques promise to significantly advance the state of the art in computer vision and to improve our understanding of perceptual inference in general.

In addition to improved performance, the approach to shape from shading described above offers a new degree of flexibility that should allow shading to be exploited in more general and realistic scenarios. Previous approaches to shape from shading typically relied heavily on the exact nature of the Lambertian reflectance equations, and so could only be applied to surfaces with specific (i.e., matte) reflectance qualities with no surface markings. Also, specific lighting conditions were assumed. The approach described above applies directly to a statistical model of the relationship between shape and shading, and so it does not depend on the exact nature of the Lambertian equation or specific lighting arrangements. Also, the efficient higher-order belief propagation techniques described here make it possible to exploit stronger, nonpairwise models of the prior probability of 3D shapes. Because the problem of depth inference is so highly underconstrained, and natural images admit large numbers of plausible 3D interpretations, it is crucial to utilize an accurate model of the prior probability of 3D surface. Knowing what 3D shapes commonly occur in nature and what shapes are *a priori* unlikely or odd is a very important constraint for depth inference. Finally, the factor graph representation of the shape-from-shading problem (see Figure 1.6) can be generalized naturally to exploit other depth cues, such as occlusion contours, texture, perspective, or the da Vinci correlation and shadow cues. The state-of-the-art approaches to the inference of depth from binocular stereo pairs typically employ belief propagation over a MRF. These approaches can be combined with our shape-from-shading framework in a fairly straightforward way, allowing both shading and stereo cues to be simultaneously utilized in a statistically optimal way. Statistical approaches to depth inference make it possible to work toward a more unified and robust depth inference framework, which is likely to become a major area of future vision research.

## ACKNOWLEDGMENTS

This work was supported by NSF CISE IIS 0713206 and AFOSR FA9550-09-1-0678 to T. S. Lee, and an NSF Graduate Research Fellowship to B. Potetz.

# CHAPTER 2

# Active Segmentation: A New Approach

Ajay Mishra
Yiannis Aloimonos

## CONTENTS

## 2.1 INTRODUCTION

Segmenting a scene (or image) into regions is an important step in visual processing. The regions are more discriminative than the individual pixels and fewer in number than the total number of pixels. This makes regions better suited and computationally less expensive to use for high-level visual processing like tracking, recognizing objects, and three-dimensional (3D) reconstruction. But to make segmentation an essential first step of vision algorithms, the segmentation algorithm needs to be consistent in its output and should not require any user input.

Let us give a state-of-the-art example to explain what it means for a segmentation algorithm to be fully automatic and consistent. We consider only a single image here shown in Figure 2.1a, but our analysis is not restricted to single images only. Most segmentation algorithms amount to statistical

(a)

(b)

(c)

FIGURE 2.1 **(See color insert.)** Segmentation results by Jianbo Shi and Jitendra Malik (2000) for the image (a) for the two cases when a number of regions are chosen to be 10 and 60, as shown in (b) and (c), respectively. If the trees are the object of interest, the segmentation in (b) is suitable, whereas if the horse is of interest, then the segmentation given in (c) is more suitable.

optimization procedures that require an estimate of the number of regions either directly or indirectly (in terms of other thresholds) to segment the scene. Figure 2.1b,c are the segmentation of Figure 2.1a using (Shi and Malik, 2000) with its parameter (number of regions) set to 10 and 60, respectively. If the trees in the scene are of interest, Figure 2.1b is more meaningful than Figure 2.1c, wherein the tree has been oversegmented into a large number of small regions. Such segmentation is termed *oversegmentation* because the region of interest is split into many small regions. However, if the tiny horse in Figure 2.1a is of interest, then Figure 2.1c is more meaningful, as the horse appears as a combination of just two regions in the image. The segmentation in Figure 2.1b, on the other hand, does not even have the region corresponding to the horse. It is a case of undersegmentation, wherein small regions are merged to form bigger regions. Depending upon which is an important region and at what scale it exists, the appropriate parameter needs to be set for (Shi and Malik, 2000) to output a meaningful segmentation.

But, in a real scene, there can be multiple regions of interest in the scene, and they can exist at dramatically different scales, as is the case in this example. So, a consistent and automatic segmentation algorithm is that which can segment the region of interest irrespective of its size in the image.

Such a segmentation is associated with the region of interest. Unlike the current algorithms, it will only segment the region of interest as foreground and everything else in the scene as background. In fact, we know that the human (primate) visual system works in a similar way. It observes and makes sense of a dynamic scene (video) or static scene (image) by making a series of fixations at various salient locations in the scene. Researchers have studied in great length about where the human eye fixates (Cref et al., 2008; Walther and Koch, 2006), but little is known about the operations carried out in the human visual system during a fixation. It is also well known that the structure of the human retina is such that only the small neighborhood around these fixations is captured in high resolution by the fovea, while the rest of the scene is captured in lower resolution by the sensors on the periphery of retina. We suggest that, as a first step, the visual system segments the region on which it is fixating. Rather than segmenting the entire scene or image at once, we suggest the human visual system segments the scene in terms of individual regions for each of the fixations in the scene. For example, as the attention of the visual system is drawn to the horse (Figure 2.2a), it segments the horse (Figure 2.2b). When the attention is drawn to the trees (Figure 2.2c), it segments the trees (Figure 2.2d).

(a)

(b)

(c)

(d)

FIGURE 2.2 For two different fixations shown in (a) and (c) by an "X,"
the corresponding regions segmented by our fixation-based algorithm are
shown in (b) and (d), respectively. Note that only the region corresponding
to the fixation is segmented.

Another important factor in designing a segmentation algorithm is how
it uses the visual cues to find the region. There are two types of visual cues:
monocular cues (namely, intensity, color, and texture) that come from a
single image only, binocular cues (such as depth), and motion cues that are
calculated by using two or more images of the scene. Usually, the existing
segmentation algorithms handle either of these two types of cues. Most
importantly, the basic framework changes significantly from the image
segmentation algorithms that use only monocular cues to the motion
segmentation algorithms that use motion cues to segment the region in
the image. Instead, the framework of the desired segmentation algorithm
should be cue independent, meaning the method to find the region should
not change if we just have monocular cues to the case when color, texture,
depth, and motion cues are all available.

In this chapter, we formulate the problem of segmentation such that it incorporates all available cues to segment the region of interest being fixated at. Our

segmentation procedure is a two-step process: first, all available visual cues are used to generate a probabilistic boundary edge map. The grayscale value of an edge pixel in the map is proportional to the probability of that pixel to be at a region boundary. In Section 2.3.1, the method to obtain this map is explained in detail. Second, the fixation point is selected in the scene either by a visual attention module or by any other meaningful strategy. The probabilistic edge map from the first step is then transferred from the Cartesian space to the polar space with the fixation point as its pole. The closed boundary of the region containing the fixation point in the Cartesian space is found as the optimal cut through the polar probabilistic edge map as described in Section 2.3.2. Splitting the segmentation framework into two steps is important, because this way the visual cues are used only to obtain a better probabilistic boundary edge map. Once that is calculated, the segmentation is defined optimally for every fixation selected in the scene (or image).

The rest of the chapter is organized as follows: in the next section, the existing segmentation algorithms are discussed in detail. In Section 2.3, our segmentation algorithm is described. The experimental results are presented with quantitative analysis in Section 2.4. We discuss the fixation strategy and how the segmentation output is stable against the location of fixation inside the region in Section 2.5. We conclude our chapter with some suggestions for future research in this area.

## 2.2 BACKGROUND

Without prior knowledge or context and only on the basis of signal processing, whatever the segmentation algorithm may be, somehow it needs to decide when to stop growing the segments and stop the process of segmentation. And that input comes from the user. Without any user input, segmenting an image into regions is an ill-posed problem, because segmentation can be fine or coarse depending on when the process is stopped. Most popular algorithms amount to such global methods. It is also widely known that it is difficult to estimate the input parameters automatically for any given image.

Several interactive algorithms have been proposed where the objective is to always segment the entire image into two regions: foreground and background. There are different types of these algorithms, and they take inputs from the user differently. These algorithms are not automatic and cannot be used to build an autonomous visual system. They are used in interactive applications, such as image/video editing and image databases.

Segmentation approaches can be broadly classified into two main categories: image segmentation where monocular cues are used to segment the image, and motion segmentation where motion cues are used to segment the image. Here, we provide a brief overview of both types of segmentation algorithms.

## 2.2.1 Image Segmentation

An image is a two-dimensional array of pixels where every pixel has a color, intensity, and texture information. A region is a connected set of pixels in the image that have similar color, intensity, and texture information. These regions are either obtained by clustering the pixels into coherent groups (such methods are called *region-based methods*) or by identifying the closed boundaries along the edges in the image formed by the gradients in color, intensity, and textural values. The closed boundaries are the closed paths through the gradient map of the monocular cues in the image. Each of these closed contours corresponds to a region.

### 2.2.1.1 Region-Based Methods

An image is considered to be a graph with each pixel represented by a node in the graph that is connected to the neighboring pixels. The edge connecting two pixels $i$ and $j$ is weighted according to the features of the pixels. In (Malik et al., 2001), the edge weight is computed based on the texture cues and the intervening contour between the pixels. The graph is divided into clusters using eigenvectors of the similarity matrix formed by collecting all the edge weights. In (Felzenszwalb and Huttenlocher, 2004), the dissimilarity of the color information of the pixels is used to assign the weights to the edges, and clusters are formed in an hierarchical clustering fashion. The criterion to group the pixels at different levels of the hierarchy is adapted to the degree of variability among clusters at that level. The algorithms in (Felzenszwalb and Huttenlocher, 2004; Fowlkes et al., 2007) and all other region-based segmentation algorithms need user input to stop the process of grouping the pixels. The algorithm in (Malik et al., 2001) needs the expected number of regions as input, whereas (Felzenszwalb and Huttenlocher, 2004) takes the threshold to stop the clustering process. In fact, without any user input, it is impossible to define the optimal segmentation. There are many other segmentation algorithms (Tu and Zhu, 2002; Zahn, 1971) based on global user parameters, like the number of regions or threshold.

Unlike the global parameter–based segmentation algorithms, the interactive segmentation algorithms (Bagon et al., 2008; Boykov and Jolly, 2001;

Rother et al., 2004; Yu and Shi, 2001) always segment the entire image into only two regions: foreground and background. The algorithm in (Boykov and Jolly, 2001) poses the problem of foreground/background segmentation as a binary labeling problem that is solved exactly using the maxflow algorithm (Boykov and Kolmogorov, 2004). It, however, requires users to label some pixels as foreground or background to build their color models. The algorithm in (Blake et al., 2004) improved upon (Boykov and Jolly, 2001) by using a Gaussian mixture Markov random field to better learn the foreground and background models. The algorithm in (Rother et al., 2004) requires users to specify a bounding box containing the foreground object. A seed point for every region in the image is required in (Arbelaez and Cohen, 2008). For foreground/background segmentation, at least two seed points are needed. Although these approaches give impressive results, they cannot be used as an automatic segmentation algorithm as they critically depend upon the user inputs. In (Yu and Shi, 2001), the algorithm tries to automatically select the seed points by using spatial attention-based methods and then uses these seed points to introduce extra constraints into their normalized cut-based formulation.

The algorithms in (Bagon et al., 2008; Veksler, 2008) need only a single seed point from the user. That in (Veksler, 2008) imposes a constraint on the shape of the object to be a star, meaning the algorithm prefers to segment the convex objects. Also, the user input for this algorithm is critical, as it requires the user to specify the center of the star shape exactly in the image. The algorithm in (Bagon et al., 2008) needs only one seed point to be specified on the region of interest and segments the foreground region using a compositional framework. But the algorithm is computationally intensive. It runs multiple iterations to arrive at the final segmentation.

### 2.2.1.2 Contour-Based Methods

Contour-based segmentation methods start with finding edge fragments in the image first, and then joining the edge fragments to form closed contours. The regions are enclosed by each of these closed contours. Due to the presence of textures and low contrast regions in the image, detecting edge fragments is a problem. The second step of joining the edge fragments is done in probabilistic fashion using image statistics. In (Williams and Jacobs, 1997), the first-order Markov model is used for contour shape, and the contours were completed using random walk. In (Ren and Malik, 2002), multiple scales are used to join the contours using orientation and texture cues. The algorithms in (Ren and Malik, 2002; Sinop and Grady, 2007; Yu and Shi, 2001) are edge-based segmentation methods.

Similar to the global region-based segmentation methods, edge-based segmentation algorithms suffer from the ambiguity of choosing the appropriate closed loops that are actually the boundaries of the regions in the image. To avoid that confusion, the contour-based interactive segmentation algorithms (Barrett and Mortensen, 1997; Mortensen and Barrett, 1995) need the user to specify the seed points along the contour to be traced. The algorithms in (Kass et al., 1988; Xu and Prince, 1998) need the user to initialize a closed contour that then evolves to adjust the actual boundary in the image.

### 2.2.2 Motion Segmentation

Prior research in motion segmentation can broadly be classified into two groups:

a. The approaches relying on two-dimensional (2D) motion measurements (Bober and Kittler, 1994; Burt et al., 1989; Odobez and Bouthemy, 1995; Weiss, 1997). There are many limitations in these techniques. Depth discontinuities and independently moving objects both cause discontinuities in the 2D optical flow, and it is not possible to separate these factors without 3D motion and structure estimation. Generally, dense optical flow is calculated at every point in the image, and like in the image segmentation, the flow value of each pixel is used to decide similarity between the pixels that are used to cluster them into regions with consistent motion. The main problem with this approach is that the optical flow is inaccurate at the boundaries; hence, the region obtained by this approach has generally poor boundaries.

   To overcome this problem, many algorithms first segment the frames into regions and then merge the regions by comparing the overall flow of the two regions. The accuracy of this method is dependent on the accuracy of the image segmentation step. If a region is produced by the image segmentation step that includes parts from different objects in the scene, it cannot be corrected by the later processing of combining regions into bigger regions. To avoid that problem, some techniques oversegment the image into small regions to reduce the chances of having overlapping regions.

But, discriminating small regions on the basis of their overall flow is difficult.

b. 3D approaches that identify clusters with consistent 3D motion (Adiv, 1985; Costeira and Kanade, 1995; Nelson, 1991; Sinclair, 1993; Thompson and Pong, 1990; Torr and Murray, 1994; Zhang et al., 1988) using a variety of techniques. Some techniques, such as (Weber and Malik, 1997), are based on alternate models of image formation. These additional constraints can be justified for domains such as aerial imagery. In this case, the planarity of the scene allows a registration process (Ayer et al., 1994; Triggs et al., 2000; Wiles and Brady, 1995; Zheng and Chellapa, 1993), and uncompensated regions correspond to independent movement.

This idea has been extended to cope with general scenes by selecting models depending on the scene complexity (Torr et al., 1998), or by fitting multiple planes using the plane plus parallax constraint (Irini and Anadan, 1998; Sawney et al., 2000). Most techniques detect independently moving objects based on the 3D motion estimates, either explicitly or implicitly. Some utilize inconsistencies between egomotion estimates and the observed flow field, while some utilize additional information such as depth from stereo, or partial egomotion from other sensors. The central problem faced by all motion-based techniques is that, in general, it is extremely difficult to uniquely estimate 3D motion from flow. Several studies have addressed the issue of noise sensitivity in structure from motion. In particular, it is known that for a moving camera with a small field of view observing a scene with insufficient depth variation, translation and rotation are easily confused (Adiv, 1989).

## 2.3 FIXATION-BASED SEGMENTATION

Since the early attempts on Active Vision (Aloimonos et al., 1988; Bajcsy, 1988; Christensen and Eklundh, 2000), there has been a lot of work completed on problems surrounding fixation, both from computational and psychological perspectives (Pahlavan et al., 1996). Despite all this development, however, the operation of fixation never really made it into the foundations of computational vision. Specifically, the fixation point has not been a parameter in the multitude of low- and middle-level computer vision algorithms.

This is the avenue we pursue here. We reformulate a fundamental problem in computer vision—segmentation—in conjunction with fixation.

A fixation lies inside a region in the image, and all the points inside the region, including the fixation, are enclosed by the region boundary (or depth boundary). Thus, segmenting the region containing the fixation is equivalent to finding the enclosing contour, which is a connected set of boundary edge fragments in the edge map of the image, around the fixation. It appears similar to the contour-based segmentation methods where they try to connect the contours into a closed boundary. But, our fixation-based algorithm is different in two ways: first, we are concerned with only the optimal contour enclosing the region being fixated at; second, the closed contour is found in the polar coordinate system rather than in the Cartesian space (used by (Williamo and Jacobs, 1997)), making the process of finding the optimal contour unaffected by the scale of the region being segmented (explained with an example later). Also, the contour-based completion algorithms focus only on the orientation of the edge fragments. We, in addition to the orientation, use the strength of the edge pixels to find the closed contour around the fixation. The strength of the edge pixel represents the probability of that edge pixel lying on a region boundary in the image. (Such an edge map is called a *probabilistic boundary edge map*.)

We are now going to explain how we use different cues to obtain the probabilistic boundary edge map. We also explain the rationale for using polar space instead of the Cartesian space for finding the optimal contour. Then, the details of our algorithm to obtain the optimal contour around the fixation are given.

Finally, we must emphasize that we seek a physical definition of segmentation. This means that we are not simply looking for a region that contains the fixation point, we rather search for a surface that contains the fixation point. In other words, the closed boundary surrounding the figure and separating it from the background is a depth boundary.

## 2.3.1 Computing Probabilistic Boundary Edge Map

As explained before, the probabilistic boundary edge map encodes the probability of the edge pixels to be at the region boundary as their gray value. This means the edge pixels along the boundary will be brighter than the internal or textural edges. Ideally, we would want to have a probabilistic boundary edge map wherein all bright edge pixels are the points

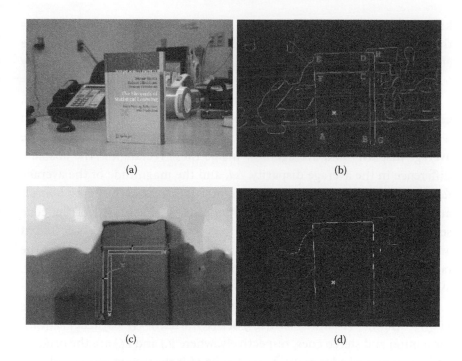

FIGURE 2.3 **(See color insert.)** (a) The first frame of a motion sequence. (b) The probabilistic boundary edge map of (a) as given by Martin, Fowlkes, and Malik, (2004). (c) The magnitude of the optical flow vectors shown in the gray image (the brightness encodes the magnitude). (d) The final boundary edge map after including motion cues.

along the region boundary (depth boundary) in the image. We are going to explore how to generate such a probabilistic boundary edge map.

Our initial probabilistic boundary edge map is the output of the Berkeley edge detector (Martin et al., 2004). Martin et al. learned the color and texture properties of the boundary pixels from the labeled data (~300 images) and used that information to differentiate the boundary edges from the internal edges. See Figure 2.3b (the edge map of Figure 2.3a) as an example of a typical output of the edge detector. Unlike binary edge detectors like Canny, it successfully removes the spurious texture edges and highlights the boundary edges, but it still has some strong internal edges (BC, CD, CF) that are not the depth boundaries.

Now, to suppress these strong internal edge segments and reinforce the boundary edges (AG, GF, FE, EA), we can use motion and stereo cues. We know that the change in disparity or flow across the internal edges is less than that across the boundary edges. So, we can look into both sides of the edges to find the change in flow and disparity and modify its probability (or gray value) accordingly.

We break the edge map into straight-line segments (such as AB, BC, CD, etc., as shown in Figure 2.3b) and select rectangular regions of width $w$ at a distance $r$ on its both sides (see Figure 2.3c). We then calculate the average disparity and average flow inside these rectangles. The absolute difference in the average disparity, $\Delta d$, and the magnitude of the average flow, $\Delta f$, are measures of how likely a segment is to be at the depth boundary. The greater the difference, the higher is the likelihood of the edge segment to be at the boundary. The rectangular regions are selected at an equal distance $r$ on both sides from the edge segment, because, at the boundary, the flow or disparity is more corrupted than inside the object. We chose $r$ and $w$ to be 5 and 10 pixels, respectively, for our experiments.

Now, the brightness of an edge pixel on the edge segment is changed as $I'(x,y) = \alpha_b I(x,y) + (1 - \alpha_b)(\Delta f/max(\Delta f))$ or $I'(x,y) + I(x,y) = (1 - \alpha_b)(\Delta d/max(\Delta d))$ for motion and stereo cues, respectively, where $I(.)$ and $I'(.)$ are the original and the improved edge maps, respectively; $\alpha b$ is the weight associated with the relative importance of the monocular cue-based boundary estimate. For our experiments, we chose $\alpha_b$ to be 0.2. The improved probabilistic boundary edge map is shown in Figure 2.3d, wherein the internal edges are dim, and the boundary edges are bright.

## 2.3.2 Why Polar?

Before we explain the method to find the optimal closed boundary around the fixation point, it is important to first explain why we choose to do so in the polar coordinate system. Let us consider finding the optimal contour for the red fixation on the disc shown in Figure 2.4a. The gradient edge map (Figure 2.4b) of the disc has two concentric circles. The big circle is the actual boundary of the disc, whereas the small circle is just the internal edge on the disc. Say that the edge map correctly assigns the boundary contour intensity 0.78 and the internal contour 0.39 (the intensity ranges from 0 to 1). The lengths of the two circles are 400 and 100 pixels, respectively. Now, the cost of tracing the boundary and the internal contour in the Cartesian space will be 88 = (400.(1 − 0.78)) and 61 = (100.(1 − 0.39)). Clearly, the internal contour costs less and hence will be considered optimal even though the boundary contour is the brightest and should actually

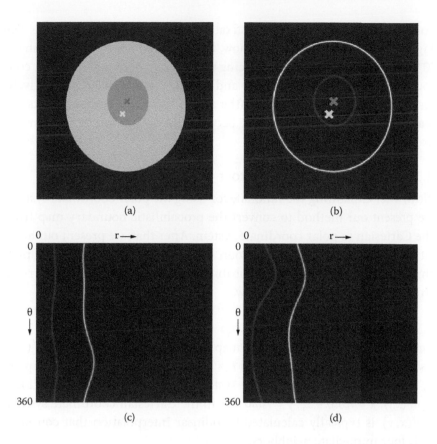

FIGURE 2.4  (a) The image of a disc. (b) Its gradient edge map. (c,d) The polar transformation of the gradient edge map in (b) with fixations shown by the Xs as its poles, respectively. In the polar representation, the radial distance increases along the horizontal axis, and angular distance increases along the vertical axis.

be the optimal contour. In fact, this problem of inherently preferring short contours over long contours has already been identified in the graph cut–based approaches where the minimum cut usually prefers to take a "short cut" in the image (Sinop and Grady, 2007).

To fix this "short-cut" problem, we have to transfer these contours to a space where their lengths no longer depend upon the area they enclose in the Cartesian space. And, the cost of tracing these contours in this space will now be independent of their scales in the Cartesian space. The polar space has this property, and we use it to solve the scale problem. The contours are transformed from the Cartesian coordinate system to the polar coordinate system with the red fixation in Figure 2.4b as the pole. See Figure 2.4c. In the polar space now, both contours become open curves (0° to 360°). Thus, the costs of tracing the inner contour and the outer contour become 80.3 = 365(1 − 0.78) and 220.21 = 361(1 − 0.39), respectively. As expected, the outer contour (the actual boundary contour) costs the least in the polar space and hence becomes the optimal enclosing contour around the fixation.

### 2.3.3 Finding the Optimal Contour

Now, after explaining the rationale for using the polar coordinate system, we present our method to convert the probabilistic boundary map from the Cartesian to polar coordinate system. After that, we present our algorithm to obtain the optimal contour, which is essentially an optimal path through the resulting polar probabilistic boundary edge map starting from its top row to its bottom row.

#### 2.3.3.1 Cartesian to Polar Edge Map

Let us say that $I_E^{cart}(.)$ is an edge map in a Cartesian coordinate, $I_E^{pol}(.)$ is its corresponding polar plot, and $F(x_o,y_o)$ is chosen as a pole. Now, a pixel $I_E^{pol}(r,\theta)$ in the polar coordinate system corresponds to a subpixel location $(x,y)$, $x = r\cos\theta + x_o$, $y = r\sin\theta + y_o$ in the Cartesian coordinate system. $I_E^{cart}(x,y)$ is typically calculated by bilinear interpolation that considers only four immediate neighbors.

We propose to generate a continuous 2D function $W(.)$ by placing 2D Gaussian kernel functions on every edge pixel. The major axis of these Gaussian kernel functions is aligned with the orientation of the edge pixel. The variance along the major axis is inversely proportional to the distance between the edge pixel and the pole $O$. Let $E$ be the set of all edge pixels. The intensity at any subpixel location $(x,y)$ in Cartesian coordinates is

$$W(x,y) = \sum_{e \in E} exp(-\frac{x_e^t}{\sigma_{x_e}^2} - \frac{y_e^t}{\sigma_{y_e}^2}) \times I^{cart}(x_e,y_e)$$

$$\begin{bmatrix} x_e^t \\ y_e^t \end{bmatrix} = \begin{bmatrix} cos\theta_e & sin\theta_e \\ -sin\theta_e & cos\theta_e \end{bmatrix} \begin{bmatrix} x_e - x \\ y_e - y \end{bmatrix},$$

where $\sigma_{x_e}^2 = K_1/\sqrt{(x_e-x_o)^2+(y_e-y_o)^2}$, $\sigma_{y_e}^2 = K_2$, $\theta e$ is the orientation at the edge pixel $e$, and $K_1 = 900$ and $K_2 = 4$ are constants. The reason for

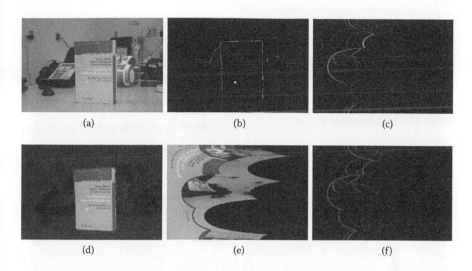

(a) (b) (c)

(d) (e) (f)

FIGURE 2.5 (a) The first frame of the image sequence captured with a moving camera. (b) The probabilistic boundary edge map. (c) The polar image of the edge map with the fixation shown by a symbol "X" in (a) and (b). (d) The segmented region containing the fixation point. (e) The image in the polar coordinate system with the fixation as the pole once again. The optimal path through the polar edge map is shown by the line. (f) The polar edge map with the optimal path through it is shown by the blue line.

setting the square of variance along the major axis, $\sigma_{x_e}^2$, to be inversely proportional to the distance of the edge pixel from the pole is to keep the gray values of the edge pixels in the polar edge map the same as the corresponding edge pixel in the Cartesian edge map. The intuition behind using variable width kernel functions for different edge pixels is as follows: Imagine an edge pixel being a finite-sized elliptical bean aligned with its orientation, and you look at it from the location chosen as a pole. The edge pixels closer to the pole (or center) will appear bigger, and those farther away from the pole will appear smaller.

The polar edge map $I_E^{pol}(r,\theta)$ is calculated by sampling $W(x,y)$. The intensity values of $I_E^{pol}$ are scaled to lie between 0 and 1. An example of this polar edge map is shown in Figure 2.5c. Our convention is that the angle $\theta \in [0°,360°]$ varies along the vertical axis of the graph and increases from the top to the bottom, whereas the radius $0 \leq r \leq r_{max}$ is represented along the horizontal axis increasing from left to right. $r_{max}$ is the maximum Euclidean distance between the fixation point and any other location on the image.

FIGURE 2.6 (a) The original image with the fixation (the "X"). (b) The probabilistic boundary edge map. (c) The segmentation based on the edge information alone. (d) The segmentation result after combining the edge information with the color information.

### 2.3.3.2 Finding the Optimal Cut through the Polar Edge
  *Map: An Inside versus Outside Segmentation*

Let us consider every pixel $p \in P$ of $I_E^{pol}$ as a node in the graph. Every pixel is connected to its four or eight immediate neighbors (Figure 2.7). The set of all the edges between nodes in the graph is denoted by $\Omega$. Let us assume $l = \{0,1\}$ are the two possible labels for each pixel where $l_p = 0$ indicates "inside" and $l_p = 1$ denotes "outside." The goal is to find a labeling $f(p)$ and $l$ that correspond to the minimum energy where the energy function is defined as

$$Q(f) = \sum_{p \in P} U_p(l_p) + \lambda \sum_{(p,q) \in \Omega} V_{p,q} . \delta(l_p, l_q)$$

$$V_{p,q} = \begin{cases} exp(-I_{E,pq}^{pol}) & \text{if } I_{E,pq}^{pol} \neq 0 \\ k & otherwise \end{cases}$$

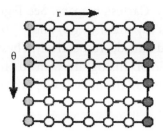 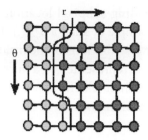

FIGURE 2.7 (Left) Initialization of the first and last columns of the polar image to be inside and outside the region of interest. (Right) The final binary labeling as a result of minimizing the energy function using graph cut.

$$\delta(l_p, l_q) = \begin{cases} 1 & \text{if } I_p \neq l_q \\ 0 & otherwise \end{cases}$$

where $\lambda = 50$, $k = 20$, $I^{pol}_{E,pq} = (I^{pol}_{E}(r_p, \theta_p) + I^{pol}_{E}(r_q, \theta_q))/2$.

At the start, there is no information about what the inside and outside of the region containing the fixation look like. So, the data term for all the nodes in the graph, except the ones in the first column and the last column, is zero ($U_p(l_p) = 0$, $\forall p \in (r, \theta)$, $0 < r < r_{max}$, $0° \Leftarrow \theta \Leftarrow 360°$). The nodes in the first column correspond to the fixation point in the Cartesian space and hence must be labeled $l_p = 0$: $U(l_p = 1) = D$ and $U(l_p = 0) = 0$ for $p \in (0, \theta)$, $0° \Leftarrow \theta \Leftarrow 360°$. The nodes in the last column must lie outside the region and are initialized to the $l_p = 1$: $U(l_p = 0)$ D and $U(l_p = 1) = 0$ for $p \in (0, \theta)$, $0° \Leftarrow \theta \Leftarrow 360°$ (see Figure 2.7). For our experiments, we chose D to be 100; the high value is to make sure the initial labels do not change as a result of minimization. We use the graph cut algorithm (Boykov and Kolmogorov, 2004) to minimize the energy function, $Q(f)$. The resulting binary segmentation is transferred back to the Cartesian space to get the desired segmentation. Figure 2.5d shows the segmentation for the fixation (the symbol "X") in the image (Figure 2.5a).

The binary segmentation as a result of the minimization step explained above splits the polar edge map into two parts: left side (inside) and right side (outside) (Figure 2.5e,f). The color information on the left (inside) and the right (outside) can now be used to modify the data term, $U_p(.)$ in the energy function $Q(f)$. The red, green, blue(RGB) value at any pixel in the polar image $I^{pol}_{rgb}(r, \theta)$ is obtained by interpolating the RGB value at the corresponding subpixel location in the Cartesian space. See Figure 2.5e for an example of such a $I_{rgb}^{pol}(.)$. Let us say that $F_{in}(r,g,b)$ and $F_{out}(r,g,b)$ are the color distributions of the inside and outside, respectively. These distributions are represented by a normalized 3D histogram with 10 bins along each color channel. The new data term for all the nodes except the first and the last column nodes is

$$U_p\big(l_p\big) = \begin{cases} \dfrac{l_n\left(F_{in}\left(I_{rgb}^{pol}\left(r_p,\theta_p\right)\right)\right)}{Z_p} & \text{if } l_p = 0 \\[4mm] \dfrac{l_n\left(F_{in}\left(I_{rgb}^{pol}\left(r_p,\theta_p\right)\right)\right)}{Z_p} & \text{if } l_p = 1 \end{cases}$$

where $Z_p = ln(F_{in}(I_p(r_p, \theta_p)) + ln(F_{out}(I^p(r_p, \theta_p))$. We again use the graph cut algorithm to minimize the energy function, $Q(f)$, with a new data term. The segmentation result improves after introducing the color information in the energy formulation. (See Figure 2.6.)

The boundary between the left (label 0) and the right (label 1) regions in the polar space will correspond to a closed contour in the Cartesian space if it is a connected path joining $(r_1, 0°)$ to $(r_1, 360°)$. But, if the two end points of this path instead are $(r_1, 0°)$ and $(r_1, 360°)$, where $r_n \neq r_1$, we assume that one of these end points is correct. We first assume $(r_1, 0°)$ and $(r_1, 360°)$ to be the correct end points. In that case, the data term is modified such that the first $r_1$ nodes from the origin in the first and the last row are assigned the label 0: $U(l_p = 1) = D$ and $U(l_p = 0) = 0$ for $p \in (r,\theta)$, $\theta \in \{0°, 360°\}$, $0 \leq r \leq r_1$. We then assume $(r_n, 0°)$ and $(r_n, 360°)$ to be the correct end points and modify the data term as $U(l_p = 1) = D$ and $U(l_p = 0) = 0$ for $p \in (r,\theta)$, $\theta \in \{0°, 360°\}$, $0 \leq r \leq r_n$. We minimize the energy functions for both cases and choose the binary segmentation corresponding to the minimum energy.

## 2.4 RESULTS

We evaluated the performance of the proposed algorithm on 20 videos with average length of seven frames and 50 stereo pairs with respect to their ground-truth segmentation. For each sequence and stereo pair, only the most prominent object of interest is identified and segmented

TABLE 2.1  Performance of Our Segmentation for the
Videos and Stereo Pairs

| For Videos | F-measure |
|---|---|
| With motion | $0.95 \pm 0.01$ |
| Without motion | $0.62 \pm 0.02$ |
| **For Stereo Pairs** | |
| With stereo | $0.96 \pm 0.02$ |
| Without stereo | $0.65 \pm 0.02$ |

*Note:* See Figure 2.8.

manually to create the ground-truth foreground and background masks. The fixation is chosen randomly anywhere on this object of interest. The videos used for the experiment are of all types: stationary scenes captured with a moving camera, dynamic scenes captured with a moving camera, and dynamic scenes captured with a stationary camera.

The segmentation output of our algorithm is compared with the ground-truth segmentation in terms of the F-measure defined as $2.P.R/(P + R)$, where $P$ stands for the precision that calculates the percentage of our segmentation overlapping with the ground truth, and $R$ stands for recall that measures the percentage of the ground-truth segmentation overlapping with our segmentation.

Table 2.1 shows that after adding motion or stereo cues with color and texture cues, the performance of the proposed method improves significantly. With color and texture cues only, the strong internal edges prevent the method from tracing the actual depth boundary. See Figure 2.8 (Row 2). However, the motion or stereo cues clean the internal edges as described in Section 2.3.1, and the proposed method finds the correct segmentation (Figure 2.8, Row 3).

To also evaluate the performance of the proposed algorithm in the presence of the monocular cues only, the images from the Alpert image database (Alpert, 2007) were used. The Berkeley edge detector (Martin et al., 2004) provides the probabilistic boundary maps of these images. The fixation on the image is chosen at the center of the bounding box around the foreground. Our definition of the segmentation for a fixation is the region enclosed by the depth boundary, which is difficult to find with the monocular cues only. Table 2.2 shows that we perform better than (Shi and Malik, 2000; Tu and Zhu, 2002) and close to (Alpert et al., 2007; Bagon et al., 2008).

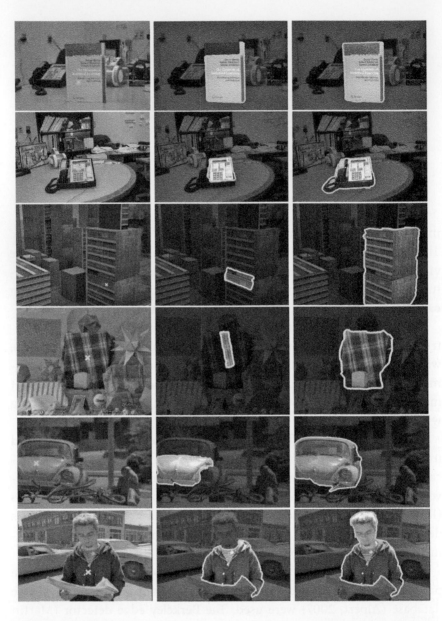

FIGURE 2.8 (Rows 1,2,3) A moving camera and stationary objects. (Row 4) An image from a stereo pair. (Row 5) A moving object (car) and a stationary camera. (Row 6) Moving objects (humans, cars) and a moving camera. (Column 1) The original images with fixations (the "X"). (Column 2) The segmentation results for the fixation using monocular cues only. (Column 3) The segmentation for the same fixation after combining motion or stereo cues with monocular cues.

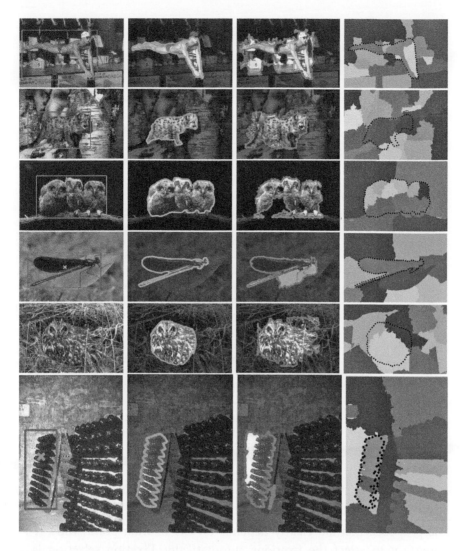

FIGURE 2.9 **(See color insert.)** The first column contains images with the fixation shown by the green "X." Our segmentation for these fixations is shown in the second column. The red rectangle around the object in the first column is the user input for the GrabCut algorithm (Rother et al., 2004). The segmentation output of the iterative GrabCut algorithm (implementation provided by www.cs.cmu.edu/~mohitg/segmentation. htm) is shown in the third column. The last column contains the output of normalized cut algorithm with the region boundary of our segmentation overlaid on it.

TABLE 2.2    Single Segment Coverage Results

| Algorithm | F-measure Score |
|---|---|
| Bagon et al. | 0.87 ± 0.010 |
| Alpert et al. | 0.86 ± 0.012 |
| Our Method | 0.83 ± 0.019 |
| NCut[a] | 0.72 ± 0.012 |
| MeanShift[b] | 0.57 ± 0.023 |

[a] See Jianbo Shi, and Jitendra Malik. *PAMI*, 22(8):888–905, 2000.

[b] See Z.W. Tu, and S.C. Zhu. *T-PAMI*, 24(5):603–619, May 2002.

*Note:* The scores for other methods except that of Shai Bagon, Oren Boiman, and Michal Irani (In David Forsyth, Philip Torr, and Andrew Zisserman, editors, *Computer Vision—ECCV 2008*, volume 5305 of *LNCS*, New York: Springer, 2008, 30–44) are taken from Sharon Alpert, Meirav Galun, Ronen Basri, and Achi Brandt, In *Proceedings of the IEEE Conference on Computer Vision and Pattern Recognition*, June 2007.

*Sources:* Shai Bagon, Oren Boiman, and Michal Irani. In David Forsyth, Philip Torr, and Andrew Zisserman, editors, *Computer Vision—ECCV 2008*, volume 5305 of *LNCS*, New York: Springer, 2008, 30–44; Sharon Alpert, Meirav Galun, Ronen Basri, and Achi Brandt. In *Proceedings of the IEEE Conference on Computer Vision and Pattern Recognition*, June 2007.

## 2.5  FIXATION STRATEGY

The proposed method clearly depends on the fixation point; thus, it is important to select the fixations automatically. Fixation selection is a mechanism that depends on the underlying task as well as other senses (like sound). In the absence of these cues, one has to concentrate on generic visual solutions. There has been a significant amount of research done on the topic of visual attention (Itti et al., 1998, Serences and Yantis, 2006; Walther and Koch, 2006), primarily to find the salient locations in the scene where the human eye may fixate. For our segmentation framework, as shown in the next section, the fixation just needs to be inside the objects in the scene. As long as this is true, the correct segmentation will be obtained. Fixation points amount to features in the scene, and the recent literature on features comes in handy (Lowe, 2004; Mikolajczyk and

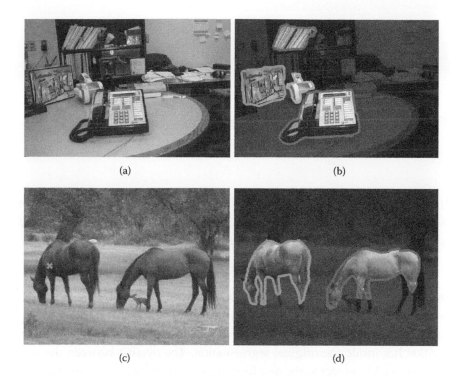

(a)                                          (b)

(c)                                          (d)

FIGURE 2.10 **(See color insert.)** (a) and (c) are the images with multiple fixations. (b) and (c) contain segmented regions corresponding to those fixations.

Schmid, 2002). Although we do not yet have a definite way to automatically select fixations, we can easily generate the potential fixations that lie inside most of the objects in a scene. Figure 2.10 shows multiple segmentation using this technique.

### 2.5.1 Stability Analysis

Here, we verify our claim that the optimal closed boundary for any fixation inside a region remains the same. The possible variation in the segmentation will occur due to the presence of bright internal edges in the probabilistic boundary edge map. To evaluate the stability of segmentation with respect to the location of fixation inside the object, we devise the following procedure: Choose a fixation roughly at the center of the object, and calculate the optimal closed boundary enclosing the segmented region. Calculate the average scale, $S_{avg}$, of the segmented region as $\sqrt{Area/\pi}$. Now, the new fixation is chosen

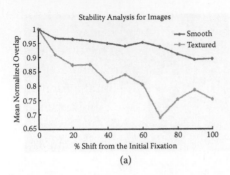
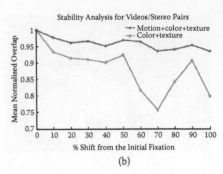

FIGURE 2.11 **(See color insert.)** Stability analysis of region segmentation with respect to the locations of fixations inside those regions for (a) the images only and (b) for videos and stereo image pairs.

by moving away from the original fixation in the random direction by $n.S_{avg}$, where $n = \{0.1, 0.2, 0.3, \ldots, 1\}$. If the new fixation lies outside the original segmentation, a new direction is chosen for the same radial shift until the new fixation lies inside the original segmentation. The overlap between the segmentation with respect to the new fixation, $R_n$, and the original segmentation, $R_o$, is given by $\frac{|R_o \cap R_n|}{|R_o \cup R_n|}$.

We calculated the overlap values for 100 textured regions and 100 smooth regions from the Berkeley Segmentation Database (BSD) and Alpert Segmentation Database. It is clear from the graph in Figure 2.11a that the overlap values are better for the smooth regions than for the textured regions. Textured regions might have strong internal edges, making it possible for the original optimal path to be modified as the fixation moves to a new location. However, for the smooth regions, there is a stable optimal path around the fixation; it does not change dramatically as the fixation moves to a new location. We also calculate the overlap values for the 100 frames from video sequences; first with their boundary edge map given by (Martin et al., 2004), and then using the enhanced boundary edge map after combining motion cues. The results are shown in Figure 2.11b. We can see that the segmentation becomes stable as motion cues suppress the internal edges and reinforce the boundary edge pixels in the boundary edge map (Martin et al., 2004).

## 2.6 CONCLUSION

We proposed here a novel formulation of segmentation in conjunction with fixation. The framework combines monocular cues with motion and stereo to disambiguate the internal edges from boundary edges. The approach is motivated by biological vision, and it may have connections to neural models developed for the problem of border ownership in segmentation (Craft et al., 2007). Although the framework was developed for an active observer, it applies to image databases as well, where the notion of fixation amounts to selecting an image point that becomes the center of the polar transformation. One of the reasons for getting good segmentation with only monocular cues is the better probabilistic boundary edge map given by (Martin et al., 2004). Our contribution here was to formulate an old problem—segmentation—in a different way and show that existing computational mechanisms in state-of-the-art computer vision are sufficient to lead us to promising automatic solutions. Our approach can be complemented in a variety of ways, for example, by introducing a multitude of cues. An interesting avenue has to do with learning models of the world. For example, if we had a model of a "horse," it could be used to get the exact contour of the horse (right) in Figure 2.10d.

In Figure 2.9, we also provide a visual comparison between the output of the proposed segmentation and the interactive Grabcut algorithm (Kass et al., 1998) and Normalized Cut (Malik et al., 2001) for some of the difficult images from the berkeley segmentation Database. For normalized cut, the best parameter (between 5 to 20) for each image is manually selected and the corresponding segmentation is shown in the last row of Figure 2.9.

# CHAPTER 3

# Mechanisms for Propagating Surface Information in 3D Reconstruction

James Coughlan

## CONTENTS

## INTRODUCTION

Bayesian and other related statistical techniques have emerged as a dominant paradigm in computer vision for estimating three-dimensional (3D) surfaces in the presence of noisy and sparse depth cues. In particular, for 3D reconstruction problems, these techniques have been implemented using Markov random fields (MRFs), which are probabilistic models that express how variables arranged in a spatial structure jointly vary, such as a rectangular grid of disparity variables aligned to a pixel lattice in a stereo model. Such MRF models incorporate powerful priors on surface geometry, which allow 3D estimates to combine evidence from depth cues with prior knowledge of smoothness constraints. They embody a natural mechanism for propagating surface information from regions with highly informative depth cues to neighboring regions with unreliable or missing depth cues, without crossing over depth discontinuities. Recent advances in inference algorithms have enlarged the range of statistical models that are tractable for computer implementation, enabling the use of increasingly realistic and expressive models and leading to some of the most robust and accurate 3D estimation algorithms to date. Finally, a "belief propagation" framework allows these models to be implemented on a massively parallel computer architecture, raising the possibility that they may be realized in a biologically plausible form.

3D reconstruction is a major theme in computer vision, with techniques for estimating shape from a variety of cues, including shading (Haines and Wilson, 2008), texture (White and Forsyth, 2006), and multiview stereo (Seitz, Curless, Diebel, Scharstein, and Szeliski, 2006). A fundamental challenge of 3D reconstruction is estimating depth information everywhere in a scene despite the fact that depth cues are noisy and sparse. These properties of depth cues mean that depth information must be propagated from regions of greater certainty to regions of lesser certainty. The 3D reconstruction problem that is perhaps the most mature in computer vision is stereo using two or more calibrated cameras with known epipolar geometry. The most successful stereo algorithms use MRFs (Chellappa and Jain, 1993), which are probabilistic models that express the joint distribution of variables arranged in a network structure, with the state of any variable exerting a direct influence on the states of its neighbors. In the case of stereo (here we consider two-view stereo), a basic MRF model consists of a lattice (grid) of disparity variables, one for each pixel in the image. For any disparity hypothesized at a given pixel, there is (usually ambiguous) evidence from the corresponding match between the left and right images that this disparity implies. Nearest-neighbor connections enforce a prior smoothness constraint on disparities, which equates to a related smoothness constraint on depths in the scene. Inference is performed using a global optimization method such as graph cuts (Boykov, Veksler, and Zabih, 2001) or belief propagation (Yedidia, Freeman, and Weiss, 2001), which determines a near-optimal assignment of disparities to each pixel given the image data. The MRF framework embodies a natural mechanism for propagating surface information in the presence of noisy and sparse data.

This chapter is intended to present the basic principles of this framework (which generalize to any 3D reconstruction problem, not just two-frame stereo) to an audience of vision researchers who are not computer vision specialists. I discuss recent extensions incorporating more realistic modeling of surface geometry, and point to recent work suggesting the possibility that belief propagation (or something close to it) may be realized in a biologically plausible form. Finally, I suggest possible directions for future research.

## MARKOV RANDOM FIELDS FOR STEREO

In this section, we outline a simple MRF (Markov random field) formulation of stereo, which we describe using a Bayesian model. (Other statistical variants such as conditional random fields (Lafferty, McCallum, and Pereira, 2001; Scharstein and Pal, 2005), which we mention below, are popular alternatives that are similar.) We are given two grayscale images $L$ and $R$ (left and right), which are assumed rectified so that a pixel in one image is guaranteed to match a pixel in the same row in the other image. The unknown disparity field is represented by $D$, with $D_r$ representing the disparity at pixel location $r$. A particular disparity value $D_r$, where $r = (x, y)$, has the following interpretation: $(x + D_r, y)$ in the left image corresponds to $(x, y)$ in the right image.

We define a prior on the disparity field $D$ that enforces smoothness:

$$P(D) = \frac{1}{Z} e^{-\beta V(D)}$$

(3.1)

where $Z$ is a normalizing constant ensuring that $P(D)$ sums to 1 over all possible values of $D$, $\beta$ is a positive constant that controls the peakedness of the probability distribution (which in turn determines the importance of the prior relative to the likelihood, discussed below), and

$$V(D) = \sum_{<rs>} f(D_r, D_s)$$

(3.2)

where the sum is over all neighboring pairs of pixels $r$ and $s$. Here, $f(D_r, D_s)$ is an energy function that penalizes differences between disparities in neighboring pixels, with higher energy values corresponding to more severe penalties. (In other words, nonzero values of the first derivative of the disparity field in the $x$ and $y$ directions are penalized.) One possible choice for the function is $f(D_r, D_s) = |D_r - D_s|$. A popular variation (Zhang and Seitz, 2005) is $f(D_r, D_s) = (|D_r - D_s|, \tau)$, which ensures that the penalty can be no larger than $\tau$; this is appropriate in scenes with depth discontinuities, where a large difference between disparities on either side of a depth edge may be no less probable than a moderate difference.

Note that a prior of this form enforces a bias toward fronto-parallel surfaces, over which the disparity is constant; we will discuss ways of relaxing this bias later.

Next, we define a likelihood function, which defines how the left and right images provide evidence supporting particular disparity values:

$$P(m|D) = \prod_r P(m_r(D_r)|D_r) \tag{3.3}$$

where the product is over all pixels in the image, and $m$ is the matching error across the entire image. Specifically, $m_r(D_r)$ is the matching error between the left and right images assuming disparity $D_r$ defined as $m_r(D_r) = |L(x + D_r, y) - R(x, y)|$ (again $r = (x, y)$). (The product form assumes that the matching errors are conditionally independent given the disparity field.) A simple model for the matching error is given by

$$P(m_r(D_r)|D_r) = \frac{1}{Z'} e^{-\mu m_r(D_r)} \tag{3.4}$$

which assigns a higher penalty (lower probability) to higher matching errors.

The Bayesian formulation defines a posterior distribution of disparities given both images, given by the Bayes theorem:

$$P(D|m) = P(D)P(m|D)/P(m) \tag{3.5}$$

To perform inference with the model, one typically finds the maximum a posterior (MAP) estimate of the disparity field (i.e., the value of $D$ that maximizes the posterior).

Note that because $P(m)$ is independent of $D$, we can write the MAP estimate of $D$, denoted $D^*$, as

$$D^* = \arg\max_D P(D)P(m|D) \tag{3.6}$$

Because maximizing any function is equivalent to maximizing the log of the function, we can reexpress this as

$$D^* = \arg\max_{D}\left\{-\beta\sum_{<rs>} f(D_r,D_s) - \mu\sum_{r} m_r(D_r)\right\}$$

(3.7)

where we have removed constants independent of $D$ such as $Z$ and $Z'$. This is equivalent to

$$D^* = \arg\min_{D}\left\{\sum_{<rs>} f(D_r,D_s) + \Upsilon\sum_{r} m_r(D_r)\right\}$$

(3.8)

where $\Upsilon = \mu/\beta$ expresses the relative weight of the prior and likelihood energies.

We discuss methods for estimating the MAP in the next section.

## Model Improvements and Refinements

We note that this MRF is a particularly simple model of stereo, and that many improvements and refinements are commonly added, such as the following:

1. Considerable performance improvements have been attained by exploiting the tendency for disparity discontinuities to be accompanied by intensity edges (Scharstein and Pal, 2005). To accomplish this, the disparity smoothness function is modulated by a measure of the image gradient, so that large disparity differences between neighboring pixels are penalized less severely when there is a strong intensity difference between the pixels. (Alternatively, a general-purpose monocular segmentation algorithm may be run to determine the likely locations of edges between regions of different intensity or color, instead of relying on a purely local measure of the image gradient.) Thus, surface information is naturally propagated from regions with highly informative depth cues to neighboring regions with unreliable or missing depth cues, without crossing over depth discontinuities.

2. The matching function used in the likelihood model can be based on comparisons of image properties that are richer than grayscale intensity—for example, full red, blue, green (RGB) color, intensity

gradient (magnitude and direction), or higher-level descriptors incorporating neighboring image structure (such as DAISY) (Tola, Lepetit, and Fua, 2008).

3. MRF models may be multiscale (Felzenszwalb and Huttenlocher, 2006), with a pyramid structure coupling the original disparity lattice with subsampled (coarsened) versions of it.

4. The conditional independence assumption in Equation (3.3) can be relaxed with the use of conditional random fields, resulting in a more realistic posterior distribution.

## How MRFs Propagate Information

We now briefly explain how the MRF model propagates noisy and sparse disparity cues throughout the image. The general principle behind this process is that the prior and likelihood distributions compete for various disparity hypotheses, and the relative strength of these two sources of information automatically varies across the image according to which source is more reliable.

In Figure 3.1 we show a simple example of a scene consisting only of a flat surface oriented fronto-parallel to the cameras (so that the correct disparity field is uniform across the image, with value $d_o$). The entire surface contains a highly textured, nonperiodic pattern, except for a central region that is textureless. Everywhere in the textured part of the image, the likelihood model provides strong evidence for the correct disparity $d_o$ and very low evidence for anything but $d_o$ (in practice, the likelihood model will be less discriminating, but we assume near-perfect disparity discrimination for the sake of argument). By contrast, in the central region there will be no direct evidence favoring one disparity over another. Any attempt to perform inference using the likelihood without the prior will obviously fail in the central region. However, by incorporating the prior with the likelihood, it is easy to see that a disparity field that has a uniform value of $d_o$ over the entire image will have a higher posterior probability than any other possible disparity field, because any disparity field that deviates from $d_o$ will be less smooth than the correct disparity field (while having the same support from the likelihood model).

More realistic examples can be analyzed exhibiting similar behavior, in which the MRF prior model propagates surface information even when the disparity cues are noisy or nonexistent.

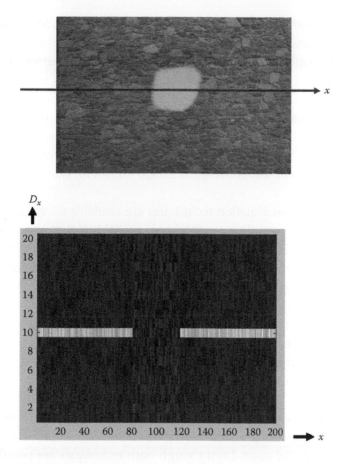

FIGURE 3.1  Idealized example of how disparity information is propagated across an image. Top: image of a fronto-parallel surface (stone wall), with x-axis slice superimposed on the image. Bottom: support for different disparities $D_x$ for all possible values of $x$, with brightness proportional to degree of support. In textured parts of a slice, the disparity value $d_0 = 10$ is strongly supported; in the textureless part of slice toward the center, all disparity values have approximately equal support. See text for explanation of how MRF model propagates $d_0 = 10$ solution across the untextured region.

## TRACTABLE INFERENCE AND LEARNING

A significant challenge posed by the stereo MRF model (and many other Bayesian models) is the difficulty in estimating the MAP, which is equivalent to minimizing the energy function in Equation (3.8). There are no techniques that are guaranteed to find the exact MAP in general cases, aside from an exhaustive search, which is impractical (e.g., $S^N$ possible disparity fields must be evaluated, where typical values are $S = 50$ for the number of possible disparities and $N = 10,000$ for the number of pixels in the image). Typical energy minimization techniques such as gradient descent (Gershenfeld, 1998) work well only when the energy function has one dominant, global minimum. However, the type of energy function defined by Equation (3.8) has a much more irregularly shaped energy landscape, with many local minima, and gradient descent-type techniques will typically converge to a local minimum that is far from the global minimum.

A variety of approximation techniques are available for estimating the MAP. Until fairly recently, one of the best available techniques was simulated annealing (Gershenfeld, 1998), which is essentially a form of a gradient descent perturbed by stochastic noise. While simulated annealing can succeed in locating a "good" local minimum (if not the global minimum), it is an extremely slow technique. In fact, its slowness meant that most MRF models in computer vision were extremely difficult to use, and as a result, progress in MRF model development was hindered.

However, approximately 10 years ago, two efficient energy minimization techniques were introduced to computer vision (imported from other fields), graph cuts (Boykov, Veksler, and Zabih, 2001), and belief propagation (Yedidia, Freeman, and Weiss, 2001). Both energy minimization techniques find approximate solutions in general, attaining exact solutions in only certain special cases. Even though both techniques are popular in a variety of computer vision MRF models, we will only discuss belief propagation (BP), which is more intuitive than graph cuts (and perhaps more related to biologically plausible mechanisms) and will provide additional insight into how MRFs propagate information. (The techniques perform minimization of energy functions with *discrete* variables, which means that the disparity values must be quantized into a finite set of values, for example, integer pixel values or subpixel values within a finite range.)

BP is a fast, iterative procedure for estimating the minimum energy (maximum probability) joint configuration of all the variables in an MRF (i.e., the joint estimate of the most likely states of all the variables). (It is also used to solve the related problem of estimating marginal probabilities of individual variables, which specify the probabilities of all possible states for each variable.) The main idea behind BP is that neighboring variables "talk" to each other at each iteration, with each variable passing "messages" to its neighbors with their estimates of the neighbors' likely states.

After enough iterations, this series of "conversations" is likely to converge to a consensus specifying which state is most likely for each variable.

The messages exchanged between variables in BP pass information throughout the MRF. Information can only be passed from variables to their immediate neighbors in one iteration of BP, but after enough iterations, it is possible for information to be passed between all variables. At the start of BP, messages express no preference for one state over another, but after enough iterations, the messages become more "opinionated," expressing strong preferences for certain states.

Finally, we note that stereo inference using BP (or graph cuts) is still slow relative to simpler (but less accurate) stereo algorithms, requiring on the order of seconds (or even minutes) for each pair of images. However, ongoing work on boosting the efficiency of these algorithms, as well as steadily improving computer hardware, is continually increasing execution speed.

## Learning MRF Parameters

The MRF model in the "Markov Random Fields for Stereo" section above has a number of free parameters (such as $\beta, \tau$, and $\mu$) that must be set correctly for the model to be realistic and accurate enough to make good inferences. There are well-established procedures (Scharstein and Pal, 2005) for learning MRF parameters from "labeled" data samples, in this case left/right image pairs and true ("ground truth") disparities. However, until recently, few datasets included ground truth disparity fields, which made it difficult to learn the MRF parameters. Fortunately, this obstacle is being removed now that there are an increasing number of datasets that include ground truth, which is determined using tools such as laser range finders (used to measure the precise depth and, hence, disparity, of nearly every pixel in a scene).

Learning MRF parameters from data not only provides a principled way of choosing the model parameters (which reflects the statistics of depth and intensity in natural scenes) but has also led to improved model performance, measured by comparing the disparity field estimated by the model with the ground truth disparities (Zhang and Seitz, 2005).

## MORE REALISTIC PRIORS

An important limitation of the MRF prior in Equation (3.1) is that it penalizes disparity differences in neighboring pixels, which implies a bias in favor of fronto-parallel surfaces. Such a bias is inappropriate for many real-world scenes with slanted surfaces. Even the toy example considered in the section on how MRFs propagate information is likely to fail if the surface is slanted: the prior may have trouble propagating the linearly changing disparity beyond the textured region of the image. In such cases, although the first $x$ and $y$ derivatives of disparity may be nonzero, the second derivatives are zero. (Any planar surface has an associated disparity field $D_r = ax + by + c$, where $r = (x, y)$, that is, the disparity is linear in the $x$ and $y$ image coordinates.)

Ongoing research in my laboratory seeks to overcome this fronto-parallel bias in the context of a specific application: terrain analysis for visually impaired wheelchair users. In this application (Ivanchenko et al., 2008), a stereo camera is pointed at the ground, such that the optical axis makes an angle of approximately 45° with the ground surface. The goal is to detect terrain irregularities such as obstacles, holes in the ground, and curbs, and to convey this information to the wheelchair user.

We designed a real-time algorithm for detecting and reporting terrain irregularities using a fast, commercially available stereo algorithm that is integrated with the stereo camera hardware. The stereo algorithm is based on simple window correlation rather than an MRF model and is therefore very fast, processing many frames per second. The disadvantage of using such a fast algorithm is that it produces sparse, noisy disparity estimates, and smooths over depth discontinuities. However, the quality of the disparity estimates suffices for detecting large terrain irregularities such as trees and other obstacles. When the algorithm fails to detect any significant deviations from the dominant ground plane (e.g., sidewalk surface) in the scene, it seems sensible to apply a more sophisticated stereo algorithm such as a MRF model to examine the scene in more detail. A second algorithm such as this may reveal the presence of a curb or other subtle depth discontinuity that was missed by the first algorithm.

The slant of the ground plane means that the disparity of the ground changes appreciably from one image row to the next, violating the fronto-parallel assumption. To rectify this problem, we are experimenting with warping one of the images so as to remove the disparity corresponding to the ground plane. (This idea was originally proposed in Burt, Wixson, and Salgian [1995]). Thus, only scene points that lie off the ground plane will have nonzero disparity, and planes parallel to the ground plane (e.g., the road bordering the sidewalk) will have *uniform* disparities. In this way, the image data are transformed so that the fronto-parallel bias is appropriate.

Such a transformation may prove valuable for our application, but a more general solution is to impose a prior that assumes that locally planar surfaces with arbitrary slant and tilt are common. One way to enforce such a prior is to penalize deviations of the second derivatives of the disparity field from zero. At a minimum, such a prior must evaluate the relationship among three consecutive pixel disparities, because a second derivative requires three consecutive samples to be estimated. (A second derivative of zero implies that the three points are collinear in 3D.) This measure is beyond the capability of the pairwise MRF presented in this chapter, and a straightforward implementation using a more powerful MRF with ternary (triplet) interactions would be extremely computationally demanding. Recent work (Woodford et al., 2008) replaces BP for such an implementation with another energy minimization algorithm that is much more efficient for this problem. The result is a tractable stereo algorithm with superior performance, particularly in its ability to propagate surface information on non-fronto-parallel surfaces.

## BIOLOGICAL PLAUSIBILITY

From time to time, I am asked by neuroscientists and psychophysicists if MRF models such as the ones described in this chapter have anything to do with biological vision systems. Although I confess to knowing little about biological vision, I would like to point to work by others arguing that the MRF-BP framework (perhaps extended to incorporate multiple depth cues) may be biologically plausible.

From a biological perspective, perhaps the most important property of models cast in this framework is that they are fully parallelizable: one can implement BP in a parallel hardware system with one computing node for each variable in the MRF, with directed connections between neighboring variables to represent BP messages. In each iteration of BP, messages flow along these connections from each variable node to neighboring nodes. Lee and Mumford (2003) have argued that BP may be a model for how information is passed top-down and bottom-up in the brain. Recent research (Ott and Stoop, 2006) has established that BP for MRFs with binary-valued variables (i.e., each variable can assume only two possible states) can be formulated with continuous time updates (rather than discrete time updates), resulting in behavior that closely matches the dynamics of a Hopfield network. Other work (Doya et al., 2006) relaxes the assumption of binary-valued variables and relates BP to a spiking network model.

## DISCUSSION

In this chapter, I presented an MRF framework for propagating surface information in 3D reconstruction in the presence of noisy and sparse depth cues. In addition to automatically weighing prior and likelihood information according to their reliability, the framework is the basis for many of the top-performing stereo algorithms in computer vision (see Scharstein and Szeliski, 2002), and the Web site associated with it, vision.middlebury. edu/stereo, which maintains up-to-date performance rankings of state-of-the-art stereo algorithms. While the standard prior used in MRF stereo algorithms imposes an unnatural fronto-parallel bias, promising recent work demonstrates the value of using a more realistic prior that accommodates the frequent occurrence of locally planar surfaces with arbitrary slant and tilt.

Although 3D reconstruction algorithms have improved a lot in recent years, much work remains. Despite the recent emphasis on learning model parameters from training data, the images used for training and testing often contain more highly textured, colorful objects than commonly occur in real-world scenes, which casts doubt on the ability of even the top-performing algorithms to generalize to the real-world domain. Additional performance measures may need to be developed to reward algorithms that minimize the kinds of catastrophic inference errors that are all too common at present, in which the disparities of some points are estimated incorrectly by tens of pixels.

More realistic priors will also be needed for algorithms to improve further. The price of increased realism may be the use of higher-level priors formulated to represent coherent surfaces, such as planar and cylindrical patches with explicit boundaries, rather than pixel-based depth or disparity fields.

Another avenue for improvement will be to integrate multiple depth cues, including monocular cues such as shading and texture, in addition to standard disparity cues. (Indeed, impressive work by Saxena, Sun, and Ng, 2007, estimates a depth field from a single color image using such cues.) It will also be important to integrate information over time (i.e., multiple video frames).

Finally, it is worth pointing out that improvements in optimization techniques such as BP will be required to realize many of the proposed extensions above, and may well influence the direction of future research.

## ACKNOWLEDGMENTS

The author was supported by National Science Foundation grant no. IIS0415310. I would like to thank Volodymyr Ivanchenko and Ender Tekin for helpful feedback on this manuscript.

Finally, it is worth pointing out that improvements in optimization techniques such as BP will be required to realize many of the proposed extensions above, and may well influence the direction of future research.

## ACKNOWLEDGMENTS

The author was supported by National Science Foundation grant no. IIS-0.5310. I would like to thank Volod, ngri rvdicheder and Blake Lebru for helpful feedback on this manuscript.

# CHAPTER 4

# 3D Surface Representation Using Ricci Flow

Wei Zeng

Feng Luo

Shing-Tung Yau

David Xianfeng Gu

## CONTENTS

## INTRODUCTION

### 3D Surface Representation

Three-dimensional (3D) surface representation plays a fundamental role in middle-level vision. In this work, a representation of 3D surfaces based on modern geometry is introduced. According to Felix Klein's Erlangen program, different geometries study the invariants under different transformation groups.

Given a surface embedded in 3D Euclidean space, $S \to \mathbb{R}^3$, intrinsically, the surface has four layers of geometric information: topology, conformal structure, Riemannian metric, and embedding, corresponding to four geometries: topology, conformal geometry, Riemannian geometry, and differential geometry for surfaces in $\mathbb{R}^3$. In order to represent the topology, genus and the number of boundaries are required; for conformal geometry, $6g - 6$ (or two) parameters are needed to describe the conformal structure of a genus $g > 1$ surface (or a torus). All genus zero closed surfaces have the same conformal structure. Given the conformal structure of $S$, a canonical conformal domain of $S$ can be uniquely determined, denoted as $D_S$; for Riemannian geometry, a function defined on the conformal domain $\lambda_S : D_S \to \mathbb{R}$ is needed to specify the Riemannian metric; and finally, by adding a mean curvature function $H_S : D_S \to \mathbb{R}$, the embedding of $S$ in $\mathbb{R}^3$ can be determined unique up to a rigid motion. Therefore, in order to represent a 3D surface, one needs a finite number of parameters to determine a canonical domain $D_S$, then two functions $\lambda, H$ defined on the domain. We denote this representation as $(D_S, \lambda_S, H_S)$, and call it conformal representation.

Converting a surface $S$ to its conformal representation $(D_S, \lambda_S, H_S)$ is a challenging problem. Ricci curvature flow offers a powerful tool for computing it.

## Ricci Curvature Flow

Intrinsic curvature flows have been used in Riemannian geometry in the past 50 years with great success. One of the most recent examples appears in the proof of the Poincaré conjecture on two manifolds (Perelman, 2002, 2003a, 2003b), where the Ricci flow is employed as a fundamental tool.

These flows deform a given Riemannian metric according to its curvature. Among the most famous ones are the Ricci flow and the Yamabe flow. Both can be used to design Riemannian metrics with special curvature properties.

The Ricci flow deforms the Riemannian metric according to its Ricci curvature. In particular, it can be used to find a metric with constant Ricci curvature. There is a simple physical intuition behind it. Given a compact manifold with a Riemannian metric, the metric induces the curvature function. If the metric is changed, the curvature will be changed accordingly. The metric can be deformed in the following way: at each point, locally scale the metric, so that the scaling factor is proportional to the curvature at the point. After the deformation, the curvature will be changed. Repeating this deformation process, both the metric and the curvature will evolve like heat diffusion. Eventually, the curvature function will become constant everywhere.

Another intrinsic curvature flow is called *Yamabe flow*. It has the same physical intuition with the Ricci flow, except for the fact that it is driven by the scalar curvature instead of Ricci curvature. For two manifolds, the Yamabe flow is essentially equivalent to the Ricci flow. But for higher-dimensional manifolds, Yamabe flow is much more flexible than the Ricci flow to reach constant-scalar-curvature metrics.

Due to the ability of intrinsic curvature flows on metric designs, two curvature flows have been recently introduced into the engineering fields: a discrete Ricci flow on surfaces and a discrete Yamabe flow on surfaces. Through these works, the power of curvature flows has been extended from pure theoretical study to solving practical problems.

One should note that in engineering fields, manifolds are usually approximated using discrete constructions, such as piecewise linear meshes; in order to employ curvature flow to solve practical problems, we need to extend the theories of curvature flows from the smooth setting to the corresponding discrete setting, and we need to pay attention to the convergence of the latter to the former. Based on the discrete theories and formula, one is allowed to design computer algorithms that can simulate and compute the flow.

## A Brief History

The theory of intrinsic curvature flows originated from differential geometry and was later introduced into the engineering fields. In this section, we give a brief overview of the literature that is directly related to the two flows mentioned above. For each flow, we would introduce some representative work on two aspects: theories in the smooth setting and the discrete setting.

### Ricci Flow on Surfaces

The Ricci flow was introduced by Richard Hamilton in a seminal paper (1982) for Riemannian manifolds of any dimension. The Ricci flow has revolutionized the study of geometry of surfaces and three-manifolds and has inspired huge research activities in geometry. In particular, it leads to a proof of the 3D Poincaré conjecture. Hamilton (1988) used the two-dimensional (2D) Ricci flow to give a proof of the *uniformization theorem* for surfaces of positive genus. This leads to a way for potential applications to computer graphics.

There are many ways to discretize smooth surfaces. The one that is particularly related to a discretization of conformality is the circle packing metric introduced by Thurston (1980). The notion of circle packing has appeared in the work of Koebe (1936). Thurston conjectured (1985) that for a discretization of the Jordan domain in the plane, the sequence of circle packings converges to the Riemann mapping. This was proved by Rodin and Sullivan (1987).

Colin de Verdière Yves (1991) established the first variational principle for circle packing and proved Thurston's existence of circle packing metrics. This paved a way for a fast algorithmic implementation of finding the circle packing metrics, such as the one by Collins and Stephenson (2003). In 2003, Chow and Luo generalized Yves' work and introduced the discrete Ricci flow and discrete Ricci energy on surfaces. They proved a general existence and convergence theorem for the discrete Ricci flow and proved that the Ricci energy is convex. The algorithmic implementation of the discrete Ricci flow was carried out by Jin et al. (2008).

Another related discretization method is called *circle pattern*; it considers both the combinatorics and the geometry of the original mesh and can be looked as a variant to circle packings. Circle pattern was proposed by Bowers and Hurdal (2003) and has been proven to be a minimizer of a convex energy by Bobenko and Springborn (2004). An efficient circle pattern algorithm was developed by Kharevych, Springborn, and Schröder (2006).

### Yamabe Flow on Surfaces

The Yamabe problem aims at finding a conformal metric with constant scalar curvature for compact Riemannian manifolds. The first proof (with flaws) was given by Yamabe (1960), which was corrected and extended to a complete proof by several researchers including Trudinger (1968), Aubin (1976), and Schoen (1984). A comprehensive survey on this topic was given by Lee and Parker (1987).

Luo (2004) studied the discrete Yamabe flow on surfaces. He introduced a notion of discrete conformal change of polyhedral metric, which plays a key role in developing the discrete Yamabe flow and the associated variational principle in the field. Based on the discrete conformal class and geometric consideration, Luo gave the discrete Yamabe energy as an integration of a differential 1-form and proved that this energy is a locally convex function. He also deduced from it that the curvature evolution of the Yamabe flow is a heat equation.

In a recent work of Springborn, Schröder, and Pinkall (2008), they were able to identify the Yamabe energy introduced by Luo with the Milnor-Lobachevsky function and the heat equation for the curvature evolution with the cotangent Laplace equation. They constructed an algorithm based on their explicit formula. Another recent work by Gu, Luo, and Yau (2009), which used the original discrete Yamabe energy from Luo (2004), has produced an equally efficient algorithm in finding the discrete conformal metrics.

The rest of the chapter is organized as follows. We first introduce some basic concepts and theories of the surface Ricci flow in the smooth setting. Then we present the discrete setting and numerical algorithms to compute the flow. In the final section, we briefly describe the applications for brain mapping, surface matching, and Teichm\"uller shape space. Further details on discrete curvature flows and their variational principles can be found in Luo (2004). The details and source codes of the algorithms presented here can be found in Gu and Yau (2008) and Gu, Luo, and Yau (2009).

## THEORIES ON THE SMOOTH SURFACE RICCI FLOW

In this section, we introduce the theory of the Ricci flow in the continuous setting, which will be extended to the discrete setting in the section below entitled "Theories on the Discrete Surface Ricci Flow."

### Fundamental Group and Universal Covering Space

The closed loops on the surface can be classified by homotopy equivalence. If two closed curves on a surface $M$ can deform to each other without leaving the surface, then they are homotopic to each other. Two closed curves sharing common points can be concatenated to form another loop. This operation defines the multiplication of homotopic classes. All the homotopy classes form the *first fundamental group* of $M$. A collection of curves on the surface is a *cut graph*, if their complement is a topological disk, which is called the *fundamental domain* of the surface.

For a genus $g$ closed surface, the fundamental group has $2g$ generators. A set of fundamental group basis $\{a_1, b_1, a_2, b_2, \ldots, a_g, b_g\}$ is *canonical*, if $a_i, b_i$ have only one geometric intersection, but neither $a_i, a_j$ nor $a_i, b_j$ have geometric intersections, where $i \neq j$. If we slice $M$ along the curves, we can get a disk-like domain with boundary $\{a_1 b_1 a_1^{-1} b_1^{-1} a_2 b_2 a_2^{-1} b_2^{-1} \ldots a_g b_g a_g^{-1} b_g^{-1}\}$, which is called the *canonical fundamental domain* of the surface $M$.

A covering space of $M$ is a surface $\bar{M}$ together with a continuous subjective map $p : \bar{M} \to M$, such that for every $q \in M$ there exists an open neighborhood $U$ of $q$ such that $p^{-1}(U)$ (the inverse image of $U$ under $p$) is a disjoint union of open sets in $\bar{M}$, each of which is mapped homeomorphically onto $U$ by $p$. If $\bar{M}$ is simply connected, then $\bar{M}$ is called the *universal covering space* of $M$. Suppose $\phi : \bar{M} \to \bar{M}, p = \phi \circ p$, then $\phi$ is called a *deck transformation*. A deck transformation maps one fundamental domain to another fundamental domain. All the deck transformations form the *deck transformation group*, which is isomorphic to the fundamental group. We use the algorithms in Carner et al. (2005) to compute the canonical fundamental group generators.

### Riemannian Metric and Gaussian Curvature

All the differential geometric concepts and the detailed explanations can be found in Guggenheimer (1977). Suppose $S$ is a $C^2$ smooth surface embedded in $\mathbb{R}^3$ with local parameter $(u_1, u_2)$. Let $r(u_1, u_2)$ be a point on $S$ and $dr = r_1 du_1 + r_2 du_2$ be the tangent vector $\mathbf{r}$ defined at that point, where

$r_1, r_2$ are the partial derivatives of $\mathbf{r}$ with respect to $u_1$ and $u_2$, respectively. The *Riemannian metric* or the *first fundamental form* is

$$< dr, dr > = \sum < r_i, r_j > du_i du_j, \quad i, j = 1, 2. \tag{4.1}$$

The Gauss map $G : S \rightarrow \mathbb{S}^2$ from the surface $S$ to the unit sphere $\mathbb{S}^2$ maps each point $p$ on the surface to its normal $\mathbf{n}(p)$ on the sphere. The *Gaussian curvature* $K(p)$ *is defined as the Jacobian of the Gauss map.* Intuitively, it is the ratio between the infinitesimal area of the Gauss image on the Gaussian sphere and the infinitesimal area on the surface.

Let $\partial S$ be the boundary of the surface $S$, $k_g$ the geodesic curvature, $dA$ the area element, $ds$ the line element, and $\chi(S)$ the Euler characteristic number of $S$. The total curvature is determined by the topology, which satisfies the following Gauss–Bonnet Theorem:

$$\int_S K dA + \int_{\partial S} k_g ds = 2\pi \chi(S). \tag{4.2}$$

Conformal Deformation

Let $S$ be a surface embedded in $\mathbb{R}^3$. $S$ has a Riemannian metric induced from the Euclidean metric of $\mathbb{R}^3$, denoted by $\mathbf{g}$. Suppose $u : S \rightarrow \mathbb{R}$ is a scalar function defined on $S$. It can be verified that $\bar{\mathbf{g}} = e^{2u}\mathbf{g}$ is also a Riemannian metric on $S$. Furthermore, angles measured by $\mathbf{g}$ are equal to those measured by $\bar{\mathbf{g}}$. Therefore, we say $\bar{\mathbf{g}}$ is a conformal deformation from $\mathbf{g}$.

A conformal deformation maps infinitesimal circles to infinitesimal circles and preserves the intersection angles among the infinitesimal circles. In Figure 4.1, we illustrate this property by approximating infinitesimal circles by finite circles. We put a regular circle packing pattern on the texture and map the texture to the surface using a conformal parameterization, where all the circles on the texture still look like circles on the surface, and all the tangency relations among the circles are preserved.

When the Riemannian metric is conformally deformed, curvatures will also be changed accordingly. Suppose $\mathbf{g}$ is changed to $\bar{\mathbf{g}} = e^{2u}\mathbf{g}$. Then, the Gaussian curvature will become

$$\bar{K} = e^{-2u}(-\Delta_g u + K) \tag{4.3}$$

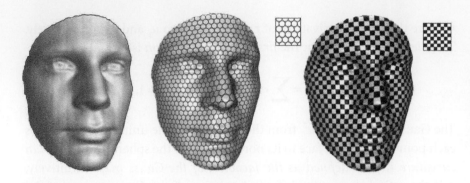

FIGURE 4.1   Properties of conformal mapping: conformal mappings transform infinitesimal circles to infinitesimal circles and preserve the inter section angles among the circles. Here, infinitesimal circles are approximated by finite ones. Notice that a circle in the texture appears in a scaled one in the texture mapping result. Also notice the angles in the checkerboard pattern preserved in the texture mapping result.

where $\Delta_g$ is the Laplacian–Beltrami operator under the original metric **g**. The geodesic curvature will become

$$\bar{\mathbf{k}} = e^{-u}(\partial_r u + \mathbf{k}) \tag{4.4}$$

where **r** is the tangent vector orthogonal to the boundary. According to the Gauss–Bonnet theorem, the total curvature is still $2\pi\chi(S)$, where $\chi(S)$ is the Euler characteristic number of $S$.

## Uniformization Theorem

Given a surface $S$ with a Riemannian metric **g**, there exist infinitely many metrics conformal to **g**. The following uniformization theorem states that among all of the conformal metrics, there exists a representative that induces constant curvature. Moreover, the constant will be one of $\{+1, 0, -1\}$.

**Theorem 4.1 (Uniformization Theorem)**

*Let $(S, \mathbf{g})$ be a compact two-dimensional surface with a Riemannian metric **g**, then there is a metric $\bar{\mathbf{g}}$ conformal to **g** with constant Gaussian curvature everywhere; the constant is one of $\{+1, 0, -1\}$. Furthermore, the constant −1 curvature metric is unique.*

We call such a metric the *uniformization metric* of $S$. According to the Gauss–Bonnet theorem (Equation [4.2]), the sign of the constant Gaussian curvature must match the sign of the Euler number of the surface: +1 for $\chi(S) > 0$, 0 for $\chi(s) = 0$, and $-1$ for $\chi(S) < 0$.

Therefore, we can embed the universal covering space of any closed surface using its uniformization metric onto one of the three canonical surfaces: the *sphere* $\mathbb{S}^2$ for genus zero surfaces with positive Euler number, the *plane* $\mathbb{E}^2$ for genus one surfaces with zero Euler number, and the *hyperbolic* space $\mathbb{H}^2$ for high genus surfaces with negative Euler number (see Figure 4.2). Accordingly, we can say that surfaces with positive Euler number admit spherical geometry; surfaces with zero Euler number admit Euclidean geometry; and surfaces with negative Euler number admit hyperbolic geometry.

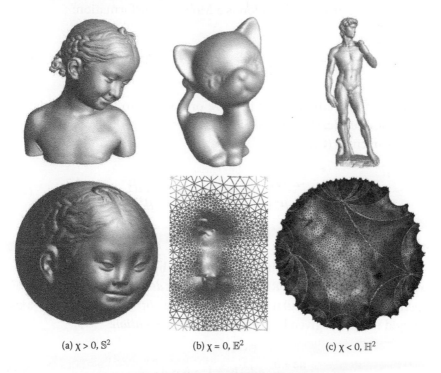

(a) $\chi > 0$, $\mathbb{S}^2$     (b) $\chi = 0$, $\mathbb{E}^2$     (c) $\chi < 0$, $\mathbb{H}^2$

FIGURE 4.2  **(See color insert.)** Uniformization theorem: each surface in $\mathbb{R}^3$ admits a uniformization metric, which is conformal to the original metric and induces constant Gaussian curvature; the constant is one of $\{+1, 0, -1\}$ depending on the Euler characteristic number $\chi$ of the surface. Its universal covering space with the uniformization metric can be isometrically embedded onto one of three canonical spaces: sphere (a), plane (b), or hyperbolic space (c). Here, we show the parameterizations computed by using discrete spherical, Euclidean, and hyperbolic Ricci flows, respectively.

## Spherical, Euclidean, and Hyperbolic Geometry

The unit sphere is with Gaussian curvature +1 and admits the spherical geometry. The rigid motions in spherical geometry are rotations. The geodesics are great arcs. The Euclidean plane is with 0 curvature and admits the Euclidean geometry. Planar translations and rotations form the rigid motion group.

The hyperbolic space model we used in this chapter is the Poincaré disk, which is a unit disk on the complex plane, with Riemannian metric

$$ds^2 = \frac{4dzd\overline{z}}{(1-z\overline{z})^2}, z \in \mathbb{C} \tag{4.5}$$

In the Poincaré disk, rigid motion is a Möbius transformation:

$$z \rightarrow e^{i\theta} \frac{z-z_0}{1-\overline{z}_0 z}, z_0 \in \mathbb{C}, \theta \in [0, 2\pi] \tag{4.6}$$

the geodesics are circular arcs that are orthogonal to the unit circle; the hyperbolic circle $(\mathbf{c}, r)$ ($\mathbf{c}$ represents the center, $r$ the radius) coincides with a Euclidean circle $(\mathbf{C}, R)$ with

$$\mathbf{C} = \frac{2-2\mu^2}{1-\mu^2 |\mathbf{c}|^2} \mathbf{c}, \ R^2 = |\mathbf{C}|^2 - \frac{|\mathbf{c}|^2 - \mu^2}{1-\mu^2 |\mathbf{c}|^{2'}} \tag{4.7}$$

where $\mu = \frac{e^r - 1}{e^r + 1}$.

We also use the upper half-plane model for hyperbolic space $\mathbb{H}^2$. $\mathbb{H}^2 = \{(x, y) \in \mathbb{R}^2 | y > 0\}$, with the Riemannian metric $ds^2 = (dx^2 + dy^2)/y^2$. In $\mathbb{H}^2$, hyperbolic lines are circular arcs and half lines orthogonal to the $x$-axis. The rigid motion is given by the *Möbius transformation*:

$$\frac{az+b}{cz+d}, ad - bc = 1, a, b, c, d \in \mathbb{R} \tag{4.8}$$

where $z = x + iy$ is the complex coordinate.

Similarly, the 3D hyperbolic space $\mathbb{H}^3$ can be represented using upper half-space model, $\mathbb{H}^3 = \{(x, y, z) \in \mathbb{R}^3 | z > 0\}$, with Riemannian metric

$$ds^2 = \frac{dx^2 + dy^2 + dz^2}{z^2} \tag{4.9}$$

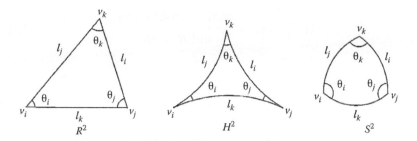

FIGURE 4.3 Euclidean, hyperbolic, and spherical triangles.

In $\mathbb{H}^3$, the hyperbolic planes are hemispheres or vertical planes, whose equators are on the $xy$-plane. The $xy$-plane represents all the infinity points in $\mathbb{H}^3$. The rigid motion in $\mathbb{H}^3$ is determined by its restriction on the $xy$-plane, which is a Möbius transformation on the plane, in the form

$$\frac{az+b}{cz+d}, ad-bc=1, a,b,c,d \in \mathbb{C} \tag{4.10}$$

Most of the computation is carried out on the $xy$-plane.

As shown in Figure 4.3, triangles with spherical, Euclidean, or hyperbolic background geometry (meaning triangles in $\mathbb{S}^2, \mathbb{E}^2,$ and $\mathbb{H}^2$) satisfy, different cosine laws:

$$(\mathbb{S}^2) \ \cos l_i = \frac{\cos\theta_i + \cos\theta_j \cos\theta_k}{\sin\theta_j \sin\theta_k} \tag{4.11}$$

$$(\mathbb{H}^2) \ \cosh l_i = \frac{\cos\theta_i + \cos\theta_j \cos\theta_k}{\sin\theta_j \sin\theta_k} \tag{4.12}$$

$$(\mathbb{E}^2) \ 1 = \frac{\cos\theta_i + \cos\theta_j \cos\theta_k}{\sin\theta_j \sin\theta_k} \tag{4.13}$$

We can interchange the role of edge and angle and get another three cosine laws:

$$(\mathbb{S}^2) \ \cos\theta_i = \frac{\cos l_i - \cos l_j \cos l_k}{\sin\theta_j \sin\theta_k} \tag{4.14}$$

$$(\mathbb{H}^2) \quad \cos\theta_i = \frac{\cosh l_i + \cosh l_j \cosh l_k}{\sinh\theta_j \sinh\theta_k} \qquad (4.15)$$

$$(\mathbb{E}^2) \quad \cos\theta_i = \frac{-l_i^2 + l_j^2 + l_k^2}{2l_j l_k} \qquad (4.16)$$

Based on the cosine laws, curvature flows on smooth surfaces can be generalized to discrete cases.

### Smooth Surface Ricci Flow

Suppose $S$ is a smooth surface with a Riemannian metric $\mathbf{g}$. The Ricci flow deforms the metric $\mathbf{g}(t)$ according to the Gaussian curvature $K(t)$ (induced by itself), where $t$ is the time parameter:

$$\frac{\mathbf{g}_{ij}(t)}{dt} = -2K(t)\mathbf{g}_{ij}(t) \qquad (4.17)$$

There is an analogy between the Ricci flow and the heat diffusion process. Suppose $T(t)$ is a temperature field on the surface. The heat diffusion equation is $dT(t)/dt = -\Delta_g T(t)$, where $\Delta_g$ is the Laplace-Beltrami operator induced by the surface metric. The temperature field becomes increasingly uniform with the increase of $t$, and it will become constant eventually.

In a physical sense, the curvature evolution induced by the Ricci flow is exactly the same as heat diffusion on the surface, as follows:

$$\frac{dK(t)}{dt} = -\Delta_{g(t)} K(t) \qquad (4.18)$$

where $\Delta_{g(t)}$ is the Laplace-Beltrami operator induced by the metric $\mathbf{g}(t)$. If we replace the metric in Equation (4.17) with $g(t) = e^{2u(t)} g(0)$, then the Ricci flow can be simplified as

$$\frac{du(t)}{dt} = -2K(t) \qquad (4.19)$$

which states that the metric should change according to the curvature. The following theorems postulate that the Ricci flow defined in Equation (4.17) is convergent and leads to a conformal uniformization metric. For surfaces with nonpositive Euler number, Hamilton (1988) proved the convergence of the Ricci flow:

### Theorem 4.2 (Hamilton, 1988)

*For a closed surface of nonpositive Euler characteristic, if the total area of the surface is preserved during the flow, the Ricci flow will converge to a metric such that the Gaussian curvature is constant everywhere.*

It is much more difficult to prove the convergence of the Ricci flow on surfaces with positive Euler number. The following result was proven by Chow (1991).

### Theorem 4.3 (Chow, 1991)

*For a closed surface of positive Euler characteristic, if the total area of the surface is preserved during the flow, the Ricci flow will converge to a metric such that the Gaussian curvature is constant everywhere.*

The corresponding metric $\mathbf{g}(\infty)$ is the uniformization metric. Moreover, at any time $t$, the metric $\mathbf{g}(t)$ is conformal to the original metric $\mathbf{g}(0)$.

The Ricci flow can be easily modified to compute a metric with a user-defined curvature $\bar{K}$ as the following:

$$\frac{du(t)}{dt} = 2(\bar{K} - K) \tag{4.20}$$

With this modification, the solution metric $\mathbf{g}(\infty)$ can be computed, which induces the curvature $\bar{K}$.

## THEORIES ON THE DISCRETE SURFACE RICCI FLOW

In engineering fields, smooth surfaces are often approximated by simplicial complexes (triangle meshes). Major concepts, such as metric, curvature, and conformal deformation, in the continuous setting can be generalized to the discrete setting. We denote a triangle mesh as $\Sigma$, a vertex set as $V$, an edge set as $E$, and a face set as $F$. $e_{ij}$ represents the edge connecting vertices $v_i$ and $v_j$, and $f_{ijk}$ denotes the face formed by $v_i$, $v_j$, and $v_k$.

### Background Geometry

In graphics, it is always assumed that a mesh $\Sigma$ is embedded in the 3D Euclidean space $\mathbb{R}^3$, and therefore each face is Euclidean. In this case, we say the mesh is with Euclidean background geometry (see Figure 4.2b). The angles and edge lengths of each face satisfy the Euclidean cosine law.

Similarly, if we assume that a mesh is embedded in the 3D sphere $\mathbb{S}^3$, then each face is a spherical triangle. We say the mesh is with spherical background geometry (see Figure 4.2a). The angles and the edge lengths of each face satisfy the spherical cosine law.

Furthermore, if we assume that a mesh is embedded in the 3D hyperbolic space $\mathbb{H}^3$, then all faces are hyperbolic triangles. We say the mesh is with hyperbolic background geometry (see Figure 4.2c). The angles and the edge lengths of each face satisfy the hyperbolic cosine law.

In the following discussion, we will explicitly specify the background geometry for a mesh when it is needed. Otherwise, the concept or the algorithm is appropriate for all kinds of background geometries.

## Discrete Riemannian Metric

A discrete Riemannian metric on a mesh $\Sigma$ is a piecewise constant metric with cone singularities. A metric on a mesh with Euclidean metric is a discrete Euclidean metric with cone singularities. Each vertex is a cone singularity. Similarly, a metric on a mesh with spherical background geometry is a discrete spherical metric with cone singularities; a metric on a mesh with hyperbolic background geometry is a discrete hyperbolic metric with cone singularities.

The edge lengths of a mesh $\Sigma$ are sufficient to define a discrete Riemannian metric:

$$l : E \to \mathbb{R}^+ \tag{4.21}$$

as long as, for each face $f_{ijk}$, the edge lengths satisfy the triangle inequality $l_{ij} + l_{jk} > l_{ki}$ for the three background geometries, and another inequality $l_{ij} + l_{jk} + l_{ki} < 2\pi$ for spherical geometry.

## Discrete Gaussian Curvature

The discrete Gaussian curvature $K_i$ on a vertex $v_i \in \Sigma$ can be computed from the angle deficit:

$$K_i = \begin{cases} 2\pi - \sum_{f_{ijk} \in F} \theta_i^{jk}, & v_i \notin \partial\Sigma \\ \pi - \sum_{f_{ijk} \in F} \theta_i^{jk}, & v_i \in \partial\Sigma \end{cases} \tag{4.22}$$

where $\theta_i^{jk}$ represents the corner angle attached to vertex $v_i$ in the face $f_{ijk}$, and $\partial\Sigma$ represents the boundary of the mesh. The discrete Gaussian curvatures are determined by the discrete metrics.

## Discrete Gauss–Bonnet Theorem

The Gauss–Bonnet theorem (Equation [4.2]) states that the total curvature is a topological invariant. It still holds on meshes as follows:

$$\sum_{v_i \in V} K_i + \lambda \sum_{f_i \in F} A_i = 2\pi\chi(M) \tag{4.23}$$

where $A_i$ denotes the area of face $f_i$, and $\lambda$ represents the constant curvature for the background geometry: +1 for the spherical geometry, 0 for the Euclidean geometry, and −1 for the hyperbolic geometry.

## Discrete Conformal Deformation

Conformal metric deformations preserve infinitesimal circles and the intersection angles among them. The discrete conformal deformation of metrics uses circles with finite radii to approximate the infinitesimal circles.

The concept of the circle packing metric was introduced in 1976 by Thurston (see Thurston, 1980) as shown in Figure 4.4. Let $\Gamma$ be a function defined on the vertices, $\Gamma : V \to \mathbb{R}^+$, which assigns a radius $\gamma_i$ to the vertex $v_i$. Similarly, let $\Phi$ be a function defined on the edges, $\Phi : E \to [0, \pi/2]$, which assigns an acute angle $\Phi(e_{ij})$ to each edge $e_{ij}$ and is called a *weight* function on the edges. Geometrically, $\Phi(e_{ij})$ is the intersection angle of two circles centered at $v_i$ and $v_j$. The pair of vertex radius function and edge weight function on a mesh $\Sigma$, $(\Gamma, \Phi)$, is called a *circle packing metric* of $\Sigma$.

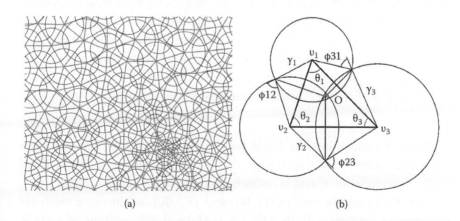

(a)    (b)

FIGURE 4.4   Circle packing metric. (a) Flat circle packing metric. (b) Circle packing metric on a triangle.

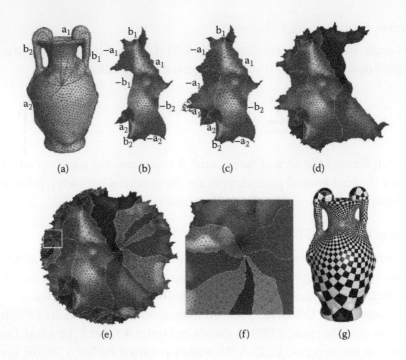

FIGURE 4.5 **(See color insert.)** The hyperbolic Ricci flow. (a) Genus two vase model marked with a set of canonical fundamental group generators, which cut the surface into a topological disk with eight sides: $a_1, b_1, a_1^{-1}, b_1^{-1}, a_2, b_2, a_2^{-1}, b_2^{-1}$. (b) The fundamental domain is conformally flattened onto the Poincaré disk with marked sides. (c) A Möbius transformation moves the side $b_1 \Rightarrow b_1^{-1}$. (d) Eight copies of the fundamental domain are glued coherently by eight Möbius transformations. (e) A finite portion of the universal covering space is flattened onto the Poincaré disk. (f) Zoom in on a region on the universal covering space, where eight fundamental domains join together. No seams or overlapping can be found. (g) Conformal parameterization induced by the hyperbolic flattening. The corner angles of checkers are well preserved.

Figure 4.4 illustrates the circle packing metrics. Each vertex $v_i$ has a circle with radius $r_i$. For each edge $e_{ij}$, the intersection angle $\phi_{ij}$ is defined by the two circles of $v_i$ and $v_j$, which either intersect or are tangent.

Two circle packing metrics $(\Gamma_1, \Phi_1)$ and $(\Gamma_2, \Phi_2)$ on the same mesh are *conformally equivalent* if $\Phi_1 = \Phi_2$. A *conformal deformation* of a circle packing metric only modifies the vertex radii and preserves the intersection angles $\Phi$ on the edges.

## Admissible Curvature Space

A mesh $\Sigma$ with edge weight $\Phi$ is called a *weighted mesh*, which is denoted as $(\Sigma, \Phi)$. In the following, we want to clarify the spaces of all possible circle packing metrics and all possible curvatures of a weighted mesh. Let the vertex set be $V = \{v_1, v_2, \ldots, v_n\}$, and the radii be $\mathbf{r} = \{r_1, r_2, \ldots, r_n\}$. Let $u_i$ be

$$
u_i = \begin{cases}
\log r_i & \mathbb{E}^2 \\[2mm]
\log \tanh \dfrac{r_i}{2} & \mathbb{H}^2 \\[2mm]
\log \tan \dfrac{r_i}{2} & \mathbb{S}^2
\end{cases}
\tag{4.24}
$$

where $\mathbb{E}^2$, $\mathbb{H}^2$, and $\mathbb{S}^2$ indicate the background geometry of the mesh. We represent a circle packing metric on $(\Sigma, \Phi)$ by a vector $\mathbf{u} = (u_1, u_2, \ldots, u_n)^T$. Similarly, we represent the Gaussian curvatures at mesh vertices by the curvature vector $\mathbf{k} = (K_1, K_2, \ldots, K_n)$. All the possible $\mathbf{u}$'s form the *admissible metric space*, and all the possible $\mathbf{k}$'s form the *admissible curvature space*.

According to the Gauss–Bonnet theory (Equation [4.23]), the total curvature must be $2\pi\chi(\Sigma)$, and therefore the curvature space is $n-1$ dimensional. We add one linear constraint to the metric vector $\mathbf{u}$, $\Sigma u_i = 0$, for the normalized metric. As a result, the metric space is also $n-1$ dimensional. If all the intersection angles are acute, then the edge lengths induced by a circle packing satisfy the triangle inequality. There is no further constraint on $\mathbf{u}$. Therefore, the admissible metric space is simply $\mathbb{R}^{n-1}$.

A curvature vector $\mathbf{k}$ is *admissible* if there exists a metric vector $\mathbf{u}$ that induces $\mathbf{k}$. The admissible curvature space of a weighted mesh $(\Sigma, \Phi)$ is a convex polytope, specified by the following theorem. The detailed proof can be found in Chow and Luo (2003).

**Theorem 4.4**

*Suppose $(\Sigma, \Phi)$ is a weighted mesh with Euclidean background geometry, I is a proper subset of vertices, $F_I$ is the set of faces whose vertices are in I, and the link set Lk(I) is formed by faces (e,v), where e is an edge and v is the third vertex in the face,*

$$
Lk(I) = \{(e, v) | e \cap I = \emptyset, v \in I\}
\tag{4.25}
$$

*then a curvature vector* **k** *is admissible if and only if*

$$\sum_{v_i \in I} K_i > - \sum_{(e,v) \in Lk(I)} (\pi - \Phi(e)) + 2\pi\chi(F_I) \qquad (4.26)$$

The admissible curvature space for weighted meshes with hyperbolic or spherical background geometries is more complicated. We refer the readers to Luo, Gu, and Dai (2007) for a detailed discussion.

Discrete Surface Ricci Flow

Suppose $(\Sigma, \Phi)$ is a weighted mesh with an initial circle packing metric. The discrete Ricci flow is defined as follows:

$$\frac{du_i(t)}{dt} = (\bar{K}_i - K_i) \qquad (4.27)$$

where $\bar{\mathbf{k}} = (\bar{K}_1, \bar{K}_2, \ldots, \bar{K}_n)^T$ is the user-defined target curvature. The discrete Ricci flow has the same form as the smooth Ricci flow (Equation [4.2]), which deforms the circle packing metric according to the Gaussian curvature, as in Equation (4.27).

The discrete Ricci flow can be formulated in the variational setting—namely, it is a negative gradient flow of a special energy form. Let $(\Sigma, \Phi)$ be a weighted mesh with spherical (Euclidean or hyperbolic) background geometry. For two arbitrary vertices $v_i$ and $v_j$, the following symmetric relation holds:

$$\frac{\partial K_i}{\partial u_j} = \frac{\partial K_j}{\partial u_i} \qquad (4.28)$$

Let $\omega = \Sigma_{i=1}^n K_i du_i$ be a differential one-form (Weitraub, 2007). The symmetric relation guarantees that the one-form is closed (curl free) in the metric space:

$$d\omega = \sum_{i,j} \left( \frac{\partial K_i}{\partial u_j} - \frac{\partial K_j}{\partial u_i} \right) du_i \wedge du_j = 0 \qquad (4.29)$$

By Stokes theorem, the following integration is path independent:

$$f(u) = \int_{u_o}^{u} \sum_{i=1}^{n} (\bar{K}_i - K_i) du \qquad (4.30)$$

where $u_0$ is an arbitrary initial metric. Therefore, the above integration is well defined and is called the *discrete Ricci energy*. The discrete Ricci flow is the negative gradient flow of the discrete Ricci energy. The discrete metric, which induces $\bar{\mathbf{k}}$, is the minimizer of the energy.

Computing the desired metric with user-defined curvature $\bar{k}$ is equivalent to minimizing the discrete Ricci energy. For Euclidean or hyperbolic cases, the discrete Ricci energy (see Equation [4.30]) was first proved to be strictly convex in the seminal work of *Yves Colin de Verdiére* (1991) for the $\Phi = 0$ case, and was generalized to all cases of $\Phi \leq \pi/2$ in Chow and Luo (2003). The global minimum uniquely exists, corresponding to the metric $\bar{u}$, which induces $\bar{k}$. The discrete Ricci flow converges to this global minimum.

**Theorem 4.5 (Chow and Luo, 2002: Euclidean Ricci Energy)**

*The Euclidean Ricci energy f(**u**) on the space of the normalized metric $\Sigma u_i = 0$ is strictly convex.*

**Theorem 4.6 (Chow and Luo, 2002: Hyperbolic Ricci Energy)**

*The hyperbolic Ricci energy is strictly convex.*

Although the spherical Ricci energy is not strictly convex, the desired metric $\bar{u}$ is still a critical point of the energy. In our experiments, the solution can be reached using Newton's method.

Figure 4.7 shows the discrete Euclidean Ricci flow for the kitten model with zero Euler number. By translation of the fundamental domain, the universal covering space of the kitten mesh is constructed, which tiles the plane. The conformality can be verified from the fact that all the corner angles in the checker-board texture are preserved. Fgiure 4.5 shows the discrete hyperbolic Ricci flow for the amphora model with negative Euler number. A finite portion of the universal covering space is flattened onto the Poincaré disk, where different periods are denoted in different colors. Similarly, the corner angles of checkers are well preserved through the conformal mapping. Figure 4.6 shows the performance of the discrete Ricci flow for different topological surfaces, such as with positive, zero, and negative Euler numbers, respectively.

## Discrete Surface Yamabe Flow

For smooth surfaces, the Ricci flow and Yamabe flow are equivalent. In the discrete case, there is a subtle difference caused by a different notion of discrete conformal classes. The following summarizes the sharp distinctions:

1. The discrete Ricci flow requires circle packing, whereas the discrete Yamabe flow is directly defined on triangulations. Therefore, Yamabe flow is more flexible.

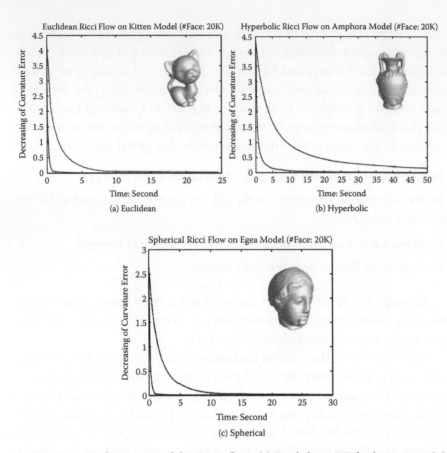

FIGURE 4.6    Performance of the Ricci flow. (a) Euclidean RF for kitten model with zero Euler number; (b) Hyperbolic RF for vase model with negative Euler number: −2; and (c) Spherical RF for human head model with positive Euler number: +2. The horizontal axis represents time; the vertical axis represents the maximal curvature error. The blue curves are for the Newton's method; the green curves are for the gradient descent method. The meshes have about 30$k$ faces. The tests were carried out on a laptop with 1.7 GHz CPU and 1 G RAM. All the algorithms are written in C++ on a Windows platform without using any other numerical library.

2. Both the Ricci flow and the Yamabe flow are variational. The energy form for the Ricci flow and the Yamabe flow are convex. But the metric space (domain of **u**) of the Ricci flow is convex, while the metric space of Yamabe flow is nonconvex. Therefore, it is stable to use Newton's method for optimizing the Ricci energy. For Yamabe energy optimization, the algorithm takes more caution.

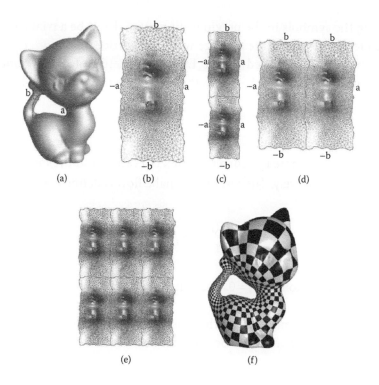

FIGURE 4.7 **(See color insert.)** The Euclidean Ricci flow. (a) Genus one kitten model marked with a set of canonical fundamental group generators $a$ and $b$. (b) A fundamental domain is conformally flattened onto the plane, marked with four sides $aba^{-1}b^{-1}$. (c) One translation moves the side $b \Rightarrow b^{-1}$. (d) The other translation moves the side $a \Rightarrow a^{-1}$. (e) The layout of the universal covering space of the kitten mesh on the plane, which tiles the plane. (f) The conformal parameterization is used for the texture mapping purpose. A checkerboard texture is placed over the parameterization in (b). The conformality can be verified from the fact that all the corner angles of the checkers are preserved.

3. Yamabe flow can adapt the connectivity to the target curvature automatically, which makes it valuable for practical purposes. During Yamabe flow, if the algorithm detects a degenerate triangle, where one angle becomes $\pi$, then the algorithm swaps the edge against the angle and continues the flow. Unfortunately, this important technique of adapting connectivity to the target curvature during the flow cannot be generalized to the Ricci flow directly.

Using the symbols in the previous discussion, let $M$ be a triangle mesh embedded in $\mathbb{R}^3$. Let $e_{ij}$ be an edge with end vertices $v_i$ and $v_j$. $d_{ij}$ is the edge length of $e_{ij}$ induced by the Euclidean metric of $\mathbb{R}^3$. A function defined on the vertices $u : V \rightarrow \mathbb{R}$ is the *discrete conformal factor*. The edge length $l_{ij}$ is defined as

$$l_{ij} = e^{u_i + u_j} d_{ij} \tag{4.31}$$

Let $K_i$ and $\bar{K}_i$ denote the current vertex curvature and the target vertex curvature, respectively. The discrete Yamabe flow is defined as

$$\frac{du_i(t)}{dt} = \bar{K}_i - K_i, \tag{4.32}$$

with initial condition $u_i(0) = 0$. The convergence of Yamabe flow is proven in Luo (2004).

Furthermore, Yamabe flow is the gradient flow of the following Yamabe energy, let $\mathbf{u} = (u_1, u_2, \ldots, u_n), n$ be the total number of vertices:

$$f(u) = \int_{u_0}^{u} \sum_{i}^{n} (\bar{K}_i - K_i) du_i \tag{4.33}$$

Similar to the Ricci flow, one can show that

$$\frac{\partial K_i}{\partial u_j} = \frac{\partial K_j}{\partial u_i} \tag{4.34}$$

The Yamabe energy is well defined and convex. The Hessian matrix can be easily constructed as follows. Supposing faces $f_{ijk}$ and $f_{jil}$ are adjacent to the edge $e_{ij}$, define the *weight* of the edge $e_{ij}$ as

$$\omega_{ij} = \cot\theta_k + \cot\theta_l \tag{4.35}$$

where $\theta_k$ is the angle at $v_k$ in $f_{ijk}$, and $\theta_l$ is the angle at $v_l$ in face $f_{jil}$. If the edge is on the boundary, and only attaches to $f_{ijk}$, then

$$\omega_{ij} = \cot\theta_k \tag{4.36}$$

It can be shown by direct computation that the differential relation between the curvature and the conformal factor is

$$dk_i = \sum_j \omega_{ij}(du_i - du_j) \qquad (4.37)$$

So the Hessian matrix of the Yamabe energy is given by

$$\frac{\partial^2 f(u)}{\partial u_i \partial u_j} = -\frac{\partial K_i}{\partial u_j} = \begin{cases} \omega_{ij}, & i \neq j \\ -\sum_k \omega_{ik}, & i = j \end{cases} \qquad (4.38)$$

The Hessian matrix is positive definite on the linear subspace $\sum_i u_i = 0$. By using the Hessian matrix formula 4.38, the Yamabe energy 4.33 can be optimized effectively. But the major difficulty is that the *admissible metric space* $\Omega(u)$ for a mesh with fixed connectivity is not convex:

$$\Omega(u) = \{u | \forall f_{ijk} \in M, l_{ij} + l_{jk} > l_{ik}\} \qquad (4.39)$$

Therefore, during the optimization process using Newton's method, we need to ensure that the metric $\mathbf{u}$ is in the admissible metric space $\Omega(u)$ at each step. If a degenerated triangle $f_{ijk}$ is detected, then we swap the longest edge of it. For example, if $\theta_k$ exceeds $\pi$, then we swap edge $e_{ij}$ as shown in Figure 4.8. The major difficulty for the discrete Ricci flow is finding a good initial circle packing with all acute edge intersection angles. This problem does not exist for discrete Yamabe flow. Therefore, Yamabe flow in general produces better conformality in practice. Figure 4.9 shows the conformal parameterizations using Yamabe flow. In frames (a) and (b), the boundary

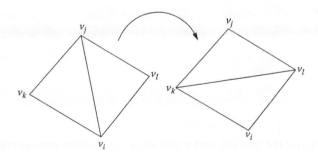

FIGURE 4.8 Edge swap.

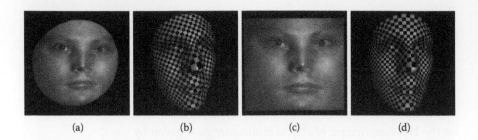

(a)                (b)                (c)                (d)

FIGURE 4.9   Conformal parameterizations using the Yamabe flow. (a, b) The boundary curvature is constant. (c, d) The curvatures at the four corners are $\pi/2$ and are zeros everywhere else.

target curvature is $2\pi/m$, where $m$ is the total number of boundary vertices. In frames (c) and (d), the curvatures at the four corners are $\pi/2$ and are zeros everywhere else. Figure 4.10 shows the hyperbolic Yamabe flow for a genus zero surface with 3 boundaries. The traget curvatures are $-1$ for interior vertices and 0 for boundary vertices. The number of edge swaps depends on the initial connectivity, initial curvatures, and target curvatures.

## APPLICATIONS

The 3D surface conformal representation using Ricci flow has broad applications.

### Conformal Brain Mapping

In medical imaging, it is helpful to compare different cortex surfaces for monitoring the progress of neurological diseases or for diagnosing the

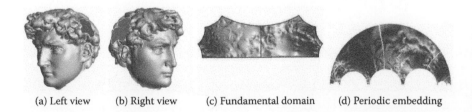

(a) Left view     (b) Right view     (c) Fundamental domain     (d) Periodic embedding

FIGURE 4.10   Hyperbolic Yamabe flow for surface with negative Euler number. (a) and (b) The left and right views of a genus zero surface with 3 boundaries. (c) The fundamental domain on the hyperbolic space $\mathbb{H}^2$. (d) The periodic conformal embedding on $\mathbb{H}^2$.

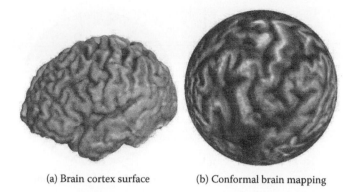

(a) Brain cortex surface          (b) Conformal brain mapping

FIGURE 4.11   Conformal brain mapping. (a) A brain cortex surface, which is a topological sphere. (b) The spherical conformal mapping of (a) onto the unit sphere.

potential abnormality. Human brain cortex surfaces are highly convoluted. It is difficult to compare the shapes of two cortical surfaces directly. By using conformal shape representation, we can map brain surfaces to the unit sphere and compare their spherical images easily. By measuring the difference between their conformal factor and mean curvature functions, the distance between the shapes can be measured quantitatively. Figure 4.11 shows an example of such conformal brain mapping. Details can be found in Gu et al. (2004).

There are prominent biological features on brain cortical surfaces. In the brain surface matching process, we can enforce the alignment of these landmarks in the following way. As shown in Figure 4.12, the cortex surface is sliced open along the landmarks and then mapped to the canonical disk with circular holes. During the matching, the corresponding boundaries are aligned together. This guarantees the matching between the corresponding landmarks.

## Surface Matching

Given two surfaces $S_1, S_2$ embedded in $\mathbb{R}^3$, we want to find a good matching $f : S_1 \rightarrow S_2$, which minimizes the distortion between them. It is generally difficult to match the surfaces directly in $\mathbb{R}^3$. As shown in Figure 4.13, we compute two conformal mappings $\phi_i : S_i \rightarrow \mathbb{D}, i = 1, 2, \mathbb{D}$ is the canonical unit disk. Then we can find a mapping $\bar{f} : \mathbb{D} \rightarrow \mathbb{D}$, which induces $f$:

$$f = \phi_2^{-1} \circ \bar{f} \circ \phi_1 \tag{4.39}$$

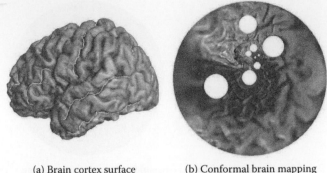

(a) Brain cortex surface        (b) Conformal brain mapping

FIGURE 4.12    Conformal brain mapping with landmarks. (a) A brain cortex surface with landmarks, which is a genus zero surface with boundaries. (b) The circular conformal mapping of (a) onto the unit disk.

For example, if we require $f$ to be isometric, then the corresponding $\bar{f}$ satisfies the following constraints: $\bar{f}$ must be conformal, and $\bar{f}$ must satisfy the following equation:

$$\lambda_1 + \mu - \lambda_2 \circ \bar{f} = 0 \tag{4.40}$$

where $e^{2\lambda_1}, e^{2\lambda_2},$ and $e^{2\mu}$ are the conformal factors of $f_1$, $f_2$, and $\bar{f}$ respectively.

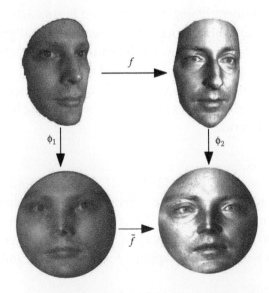

FIGURE 4.13    Surface matching using conformal geometry.

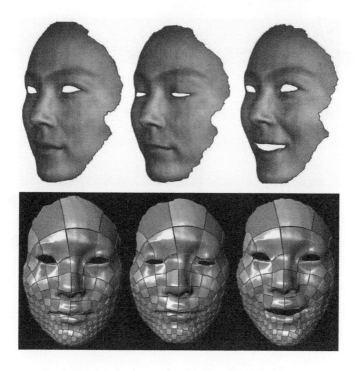

FIGURE 4.14  Surface matching using conformal geometry.

Figure 4.14 shows a matching result for a 3D human face with different expressions. The matching result is illustrated by consistent texture mapping. Details can be found in Zeng et al. (2008).

## Teichmüller Shape Space

Surfaces with the same topology can be classified according to their conformal structures. Two surfaces are conformally equivalent, if there exists a conformal mapping between them. The space of all conformal equivalence classes is called the *Teichmüller space*. According to Teichmüller theory, the Teichmüller space is also a finite dimensional Riemannian manifold.

All genus zero closed surfaces are conformally equivalent to the unit sphere; therefore, the Teichmüller space of genus zero closed surfaces has a single point. The dimension of genus one closed surfaces is two. The dimension of genus $g$ closed surfaces is $6g - 6$.

Given a surface in $\mathbb{R}^3$, the Teichmüller coordinates can be computed using the Ricci flow. The coordinates can be used for shape classification and indexing. For example, Figure 4.15 shows the Teichmüller shape space of genus zero surfaces with three boundaries. First, we compute the hyperbolic

FIGURE 4.15  **(See color insert.)** Teichmüller space of genus zero surface with three boundaries.

metric using Ricci flow, such that the boundaries become geodesics. The lengths of three boundaries under the hyperbolic form the Teichmüller coordinates.

The method can be generalized to high genus surfaces as shown in Figure 4.16. First, the hyperbolic metric is computed using Ricci flow, which is conformal to the original induced Euclidean metric. Then, there exists a unique geodesic in each homotopy class in the fundamental group. The geodesic lengths form a spectrum, which can be used as the coordinates in the Teichmüller space. Details can be found in Jin et al. (2008).

FIGURE 4.16  The geodesic spectrum under hyperbolic metric can be used to classify high genus surfaces.

## CONCLUSION AND FUTURE WORK

This work proposes a rigorous and general framework for representing 3D surfaces with arbitrary topologies in 3D Euclidean space. The geometric information of a surface is decomposed to four layers—topology, conformal structure, Riemannian metric, and mean curvature—that together determine the surface unique up to a rigid motion.

Ricci curvature flow is the essential tool to compute the surface representation. This work presents discrete curvature flow methods that are recently introduced into the engineering fields: the discrete Ricci flow and discrete Yamabe flow for surfaces. Applications for brain mapping, surface matching, and Teichmüller shape space are briefly explained.

In the future, we will study the relation between the Ricci flow and the middle-level vision.

## ACKNOWLEDGMENT

This work has been supported by NSF CCF-0448399, NSF DMS-0528363, NSF DMS-0626223, and NSF IIS-0713145.

| Color image | Range image |

Potetz FIGURE 1.2   An example color image (left) and range image (right) from our database. For purposes of illustration, the range image is shown by displaying depth as shades of gray. Notice that dark regions in the color image tend to lie in shadow, and that shadowed regions are more likely to lie slightly farther from the observer than the brightly lit outer surfaces of the rock pile. This example image from our database had an especially strong correlation between closeness and brightness.

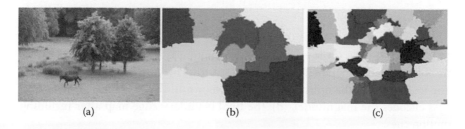

(a)          (b)          (c)

Mishra FIGURE 2.1   Segmentation results by Jianbo Shi and Jitendra Malik (2000) for the image (a) for the two cases when a number of regions are chosen to be 10 and 60, as shown in (b) and (c), respectively. If the trees are the object of interest, the segmentation in (b) is suitable, whereas if the horse is of interest, then the segmentation given in (c) is more suitable.

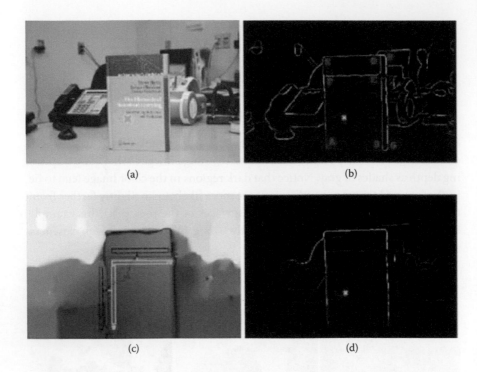

(a)          (b)

(c)          (d)

Mishra FIGURE 2.3 (a) The first frame of a motion sequence. (b) The probabilistic boundary edge map of (a) as given by Martin, Fowlkes, and Malik, (2004). (c) The magnitude of the optical flow vectors shown in the gray image (the brightness encodes the magnitude). (d) The final boundary edge map after including motion cues.

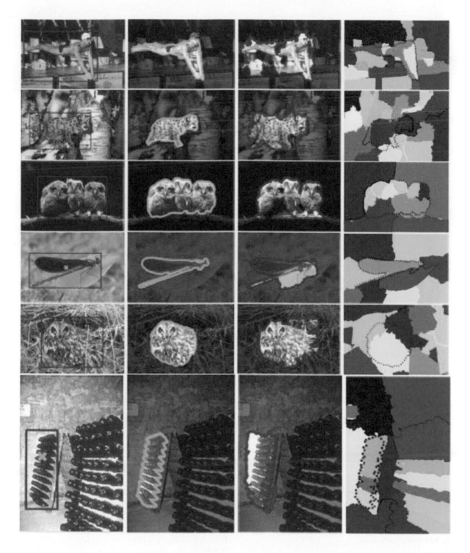

Mishra FIGURE 2.9 The first column contains images with the fixation shown by the green "X." Our segmentation for these fixations is shown in the second column. The red rectangle around the object in the first column is the user input for the GrabCut algorithm (Rother et al., 2004). The segmentation output of the iterative GrabCut algorithm (implementation provided by www.cs.cmu. edu/~mohitg/segmentation.htm) is shown in the third column. The last column contains the output of normalized cut algorithm with the region boundary of our segmentation overlaid on it.

Mishra FIGURE 2.10   (a) and (c) are the images with multiple fixations. (b) and (c) contain segmented regions corresponding to those fixations.

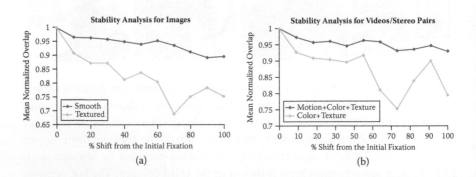

Mishra FIGURE 2.11   Stability analysis of region segmentation with respect to the locations of fixations inside those regions for (a) the images only and (b) for videos and stereo image pairs.

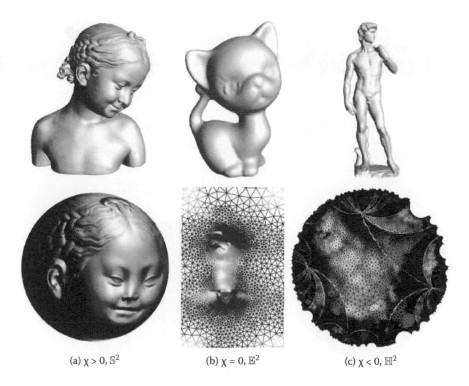

(a) $\chi > 0, \mathbb{S}^2$       (b) $\chi = 0, \mathbb{E}^2$       (c) $\chi < 0, \mathbb{H}^2$

Zeng FIGURE 4.2   Uniformization theorem: each surface in $\mathbb{R}^3$ admits a uniformization metric, which is conformal to the original metric and induces constant Gaussian curvature; the constant is one of $\{+1, 0, -1\}$ depending on the Euler characteristic number $\chi$ of the surface. Its universal covering space with the uniformization metric can be isometrically embedded onto one of three canonical spaces: sphere (a), plane (b), or hyperbolic space (c). Here, we show the parameterizations computed by using discrete spherical, Euclidean, and hyperbolic Ricci flows, respectively.

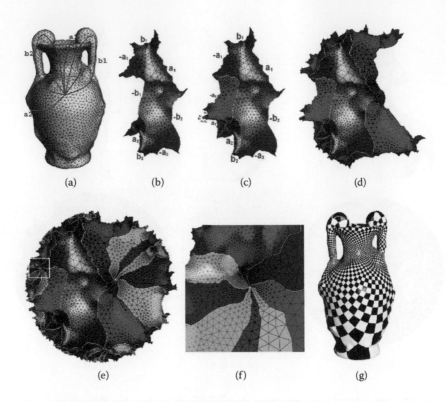

(a)          (b)          (c)          (d)

(e)                    (f)                    (g)

Zeng FIGURE 4.5   The hyperbolic Ricci flow. (a) Genus two vase model marked with a set of canonical fundamental group generators, which cut the surface into a topological disk with eight sides: $a_1, b_1, a_1^{-1}, b_1^{-1}, a_2, b_2, a_2^{-1}, b_2^{-1}$. (b) The fundamental domain is conformally flattened onto the Poincaré disk with marked sides. (c) A Möbius transformation moves the side $b_1 \Rightarrow b_1^{-1}$. (d) Eight copies of the fundamental domain are glued coherently by eight Möbius transformations. (e) A finite portion of the universal covering space is flattened onto the Poincaré disk. (f) Zoom in on a region on the universal covering space, where eight fundamental domains join together. No seams or overlapping can be found. (g) Conformal parameterization induced by the hyperbolic flattening. The corner angles of checkers are well preserved.

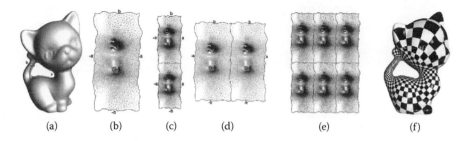

(a)     (b)     (c)     (d)     (e)     (f)

Zeng FIGURE 4.7  The Euclidean Ricci flow. (a) Genus one kitten model marked with a set of canonical fundamental group generators $a$ and $b$. (b) A fundamental domain is conformally flattened onto the plane, marked with four sides $aba^{-1}b^{-1}$. (c) One translation moves the side $b \Rightarrow b^{-1}$. (d) The other translation moves the side $a \Rightarrow a^{-1}$. (e) The layout of the universal covering space of the kitten mesh on the plane, which tiles the plane. (f) The conformal parameterization is used for the texture mapping purpose. A checkerboard texture is placed over the parameterization in (b). The conformality can be verified from the fact that all the corner angles of the checkers are preserved.

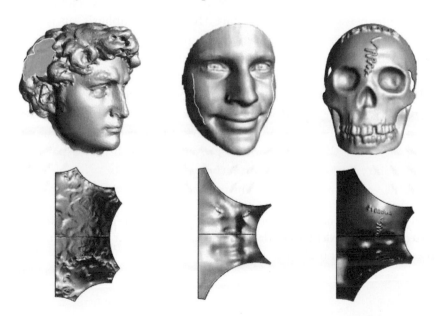

Zeng FIGURE 4.15  Teichmüller space of genus zero surface with three boundaries.

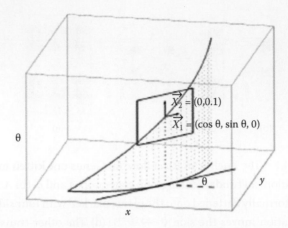

Sarti FIGURE 7.2    A two-dimensional curve (in blue) and its three-dimensional cortical lifting in the roto-translation group (in red). The tangent vector to the blue curve is (cos ($\theta$), sin($\theta$)), so that the tangent vector to its lifted curve lies in the plane generated by $\vec{X}_1$ and $\vec{X}_2$.

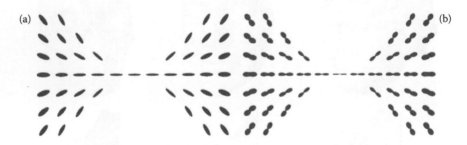

Sarti FIGURE 7.9    The fundamental solution $\Gamma$ of the Fokker–Planck equation visualized as second-order tensors (left [a]) and as infinity-order tensors by means of the density operator (right [b]).

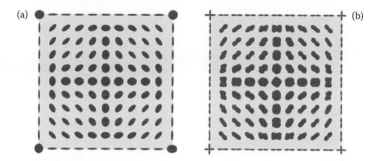

Sarti FIGURE 7.10  The inner structure of the square obtained after propagating the lifted boundaries via the Fokker–Planck fundamental solution. The probability density is visualized as second-order tensors (left [a]) and as infinity-order tensor by means of the density operator (right [b]).

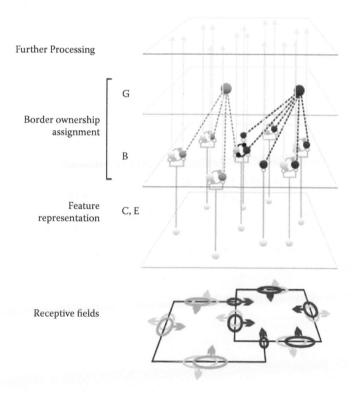

Von Der Heydt FIGURE 8.9  Neural network model of border ownership coding. See text for further explanation. Reproduced, with permission, from Craft, Schutze, Niebur, and von der Heydt (2007).

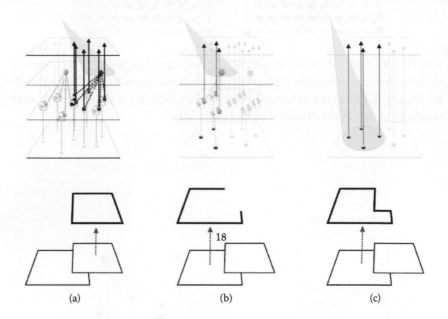

Von Der Heydt FIGURE 8.10 Explaining selective attention by the model of Figure 8.9. It is assumed that volitional ("top-down") attention excites neurons in the G cell layer as illustrated by a yellow spotlight. In this model, attention enhances the correct contour segments, whether foreground (a) or background objects (b) are attended. In contrast, a spatial attention model extracts a mixture of contours from both foreground and background objects (c). (Modified from Craft et al., 2007. With permission.)

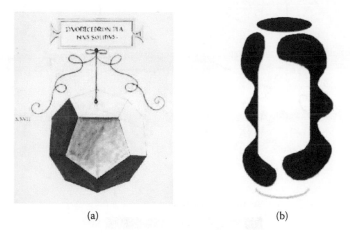

(a)                                          (b)

Tyler FIGURE 9.1    (a) Dodecahedron drawn by Leonardo da Vinci (1509). (b) Volumetric completion of the white surface of a cylinder (Reprinted with permission from Tse, 1999.) Note the strong three-dimensional percepts generated by the sparse depth cues in both images.

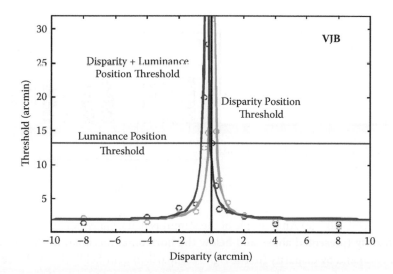

Tyler FIGURE 9.5    Typical results of the position localization task. The gray circles are the thresholds for the profile defined only by disparity; the black circles are the thresholds defined by disparity and luminance. The dashed gray line shows the model fit for disparity alone; the solid line shows that for combined disparity and luminance, as predicted by the amount of disparity required to null the perceived depth from luminance alone (Likova and Tyler, 2003). The horizontal line shows threshold for the pure luminance. Note the leftward shift of the null point in the combined luminance/disparity function.

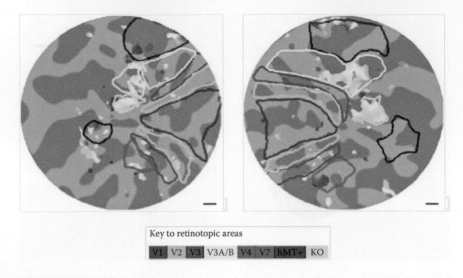

Key to retinotopic areas

V1 V2 V3 V3A/B V4 V7 hMT+ KO

Tyler FIGURE 9.7 Functional magnetic resonance imaging (fMRI) flat maps of the posterior pole of the two hemispheres showing the synchronized response to stereoscopic structure (yellowish phases) localized to V3A/B (yellow outlines) and KO (cyan outlines). (Reprinted with permission, from Tyler et al., 2006)

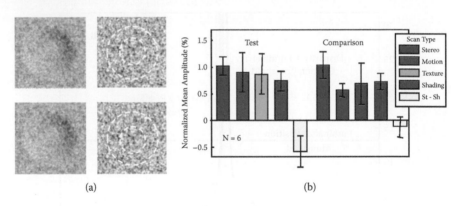

(a)                    (b)

Tyler FIGURE 9.8 (a) An example of the stimulus pairs. Gaussian bumps defined by shading (presented above and below the horizontal meridian in the left/test hemifield) and by disparity (in the right/comparison hemifield). (b) Evidence for a generic depth map in the dorsolateral occipital cortex (average of six brains). Test hemifield: Mean group cortical response to four depth cues (see key) at a dorsolateral occipital cortical location that is a candidate for the generic depth map. Note the similarity of response amplitudes for the four individual depth cues (multicolored upward bars), and no significant response in the modality alternation experiment (yellow downward bar), where disparity and shading cues are counterposed (St – Sh). Comparison hemifield: Disparity response under each condition (except the last, where the disparity was held constant as a control for the modality alternation condition). (Tyler et al., 2006)

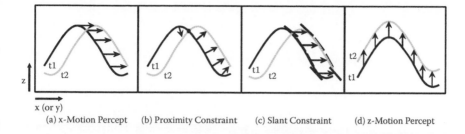

| (a) x-Motion Percept | (b) Proximity Constraint | (c) Slant Constraint | (d) z-Motion Percept |

Tyler FIGURE 9.9    (a) Diagram representing two sequential phases (full and dashed curves) of a stereoscopic sinusoidal surface (schematized as a cross-section in z,x space, i.e., a top view). Arrows show some corresponding locations, as required for the percept of lateral motion observed, while the surface waveform alternates between the two phases. (b) A proximity constraint can not explain the observed percept. (c) A surface orientation (slant) constraint would provide the requisite matches to account for the percept in (a). (d) Alternating sinusoid between near and far z-axis positions enforces a percept of z-axis stereomotion.

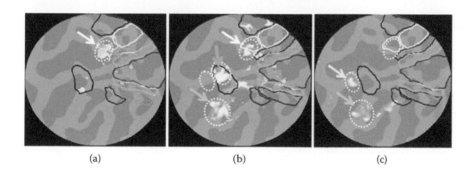

|     (a)     |     (b)     |     (c)     |

Tyler FIGURE 9.10    Occipital flat map for the left hemisphere of one subject shows distinct locations of significant activation (yellowish patches). Full-colored outlines show retinotopic areas as in Figure 9.7. Dark blue outline: boundary of hMT+ defined by a motion localizer. Dashed outlines are for comparison of clusters of activated voxels across the three conditions. (a) Stereoscopic structure of a static sinusoidal disparity versus a flat disparity plane activates a region in the dorsolateral cortex. (b) Frontoparallel ($y$-axis) stereomotion of the sinusoidal stereoscopic surface contrasted with a flat plane activates a swath of cortex including hMT+ (green arrow), together with two sites from the previous two conditions: the depth-structure region similar to (a) (white arrow) and the ventral site seen in (c) (cyan arrow). (c) $Z$-axis stereomotion versus $y$-axis stereomotion of the same stereoscopic sinusoidal surface activates regions anterior (yellow arrow; cyclopean stereomotion area CSM) and ventral (cyan arrow) to hMT+.

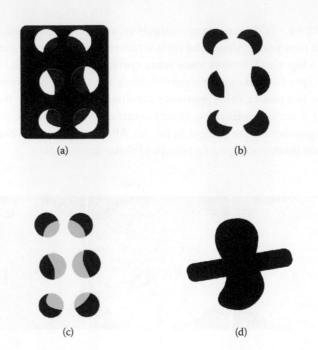

Kellman FIGURE 10.2    Four perceptual phenomena that can be explained by the same contour interpolation process. (a) A partially occluded object. The blue fragments are spatially disconnected, but we perceive them as part of the same object. (b) The same shape appears as an illusory figure and is defined by six circles with regions removed. (c) A bistable figure that can appear either as a transparent blue surface in front of six circles or an opaque blue surface seen through six circular windows. (d) A self-splitting object. The homogenous black region is divided into two shapes. This figure is bistable because the two shapes appear to reverse depth ordering over time.

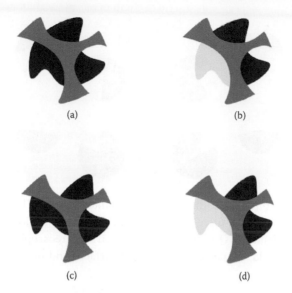

(a)                                                (b)

(c)                                                (d)

Kellman FIGURE 10.3  Contour and surface interpolation. (a) The three black
regions appear as one object behind the gray occluder. Both contour and surface
interpolation processes are engaged by this display. (b) Contour interpolation alone.
By changing the surface colors of visible regions, surface interpolation is blocked.
However, the relations of contours still engage contour interpolation, leading to
some perceived unity of the object. (c) Surface interpolation alone. By disrupting
contour relatability, contour interpolation is blocked. Due to surface interpolation,
there is still some impression that the three fragments connect behind the occluder.
(d) With both contour and surface interpolation disrupted, blue, yellow, and black
regions appear as three separate objects.

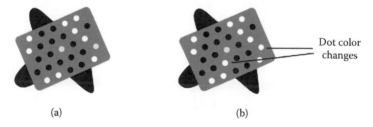

(a)                                                (b)

Kellman FIGURE 10.4    Illustration of two-dimensional surface interpolation.
The circular areas in the display do not trigger contour processes, due to the
absence of tangent discontinuities. Surface interpolation causes some circular
areas to appear as holes in the occluder rather than as spots in front. Two dots in
(a) are changed in color in (b), causing a difference in their appearance (e.g., the
yellow spot in (a) when turned white becomes a hole due to its relation with the
color of the surround). Relations of contour and surface interpolation are shown
by blue spots appearing as holes if they fall within interpolated (or extrapolated)
contours of the blue display.

Kellman FIGURE 10.7   Three-dimensional (3D) interpolation. The display is a stereogram that may be free-fused by crossing the eyes. Specification of input edges' positions and orientations in 3D space (here given by stereoscopic disparity) leads to creation of a vivid, connected, transparent surface bending in depth.

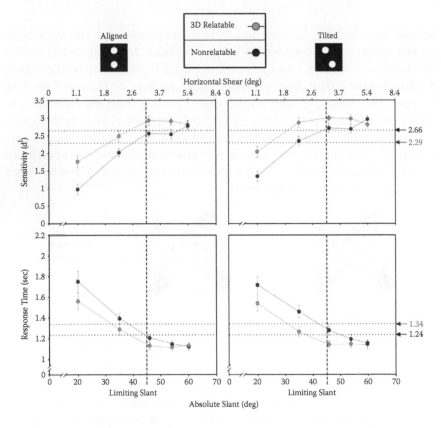

Kellman FIGURE 10.10   Three dimensional (3D) surface interpolation data. Sensitivity (upper panels) and response times (lower panels) for 3D relatable and 3D nonrelatable surface patches in aligned (right) and misaligned (left) aperture configurations. (From Fantoni et al., 2008. With permission.)

# CHAPTER 5

# Cue Interpretation and Propagation

## Flat versus Nonflat Visual Surfaces

Volodymyr V. Ivanchenko

## CONTENTS

## INTRODUCTION

In this chapter, we consider what information is available to the human visual system in cases of flat and nonflat surfaces. In the case of flat surfaces, visual cues can be interpreted without any reference to surface shape; under many conditions, cue integration is well described by a linear rule; moreover, it is possible to propagate cues along the surface to the locations where visual information is poor or missing. In the general case of nonflat surfaces, cue interpretation depends on surface shape. In such situations, the visual system can interpret surface cues in at least two different ways: shape can be ignored (i.e., a planar approximation is used instead) or surface shape can be estimated beforehand to explain away the effect of shape on interpretation of visual cues.

Numerous motor and visual tasks of everyday life require our visual system to operate with dense surface representations. Consider, for example, a simple task of placing a cup on a table. Even if a table surface has no texture and thus no visual cues for surface position and orientation, humans apparently have no difficulty in accomplishing the task. In this case, one can hypothesize that visual cues from a table border propagate across a table surface to help in the placing task. Note that situations when the visual system has to rely on sparse visual cues to build a dense surface representation are not rare. Apart from the lack of texture, a visual scene may have very few highly reliable and informative cues (e.g., originating from strong edges of regular shapes) that can be effectively used in the locations where cues are less informative. Such cue propagation should rely on prior knowledge about surface shape in order to interpolate information properly.

The most realistic scenario for cue propagation is when a visual surface is flat or slowly curving and has a known orientation. For nonflat surfaces, cue interpretation depends on surface shape. Although most of the previous research focused either on flat surfaces or surfaces with only one principle direction of curvature (e.g., cylinders), this chapter considers a more general case of nonflat surfaces. Specifically, it explores a case of cue integration when visual cues depend on both slant and shape of visual surfaces.

In the first section, I briefly review a cue integration framework that is used to estimate properties of visual scenes and visual surfaces in particular. Cue integration considers evidence from a single location, but cue propagation uses (interpolates) evidence from multiple locations to compensate for the absence or weakness of visual cues. A method for detection of weak cues and facilitation of information propagation to corresponding locations is discussed in the second section. The third section focuses on nonflat surfaces and considers a hypothesis that shape estimation precedes cue interpretation. We report the findings of two psychophysical experiments that suggest that shape cues participate in the interpretation of slant cues.

## CUE INTEGRATION

A topic of human perception and perception of visual surfaces in particular was a subject of extensive research over a period of decades. Numerous visual and motor tasks were analyzed in order to determine how visual information about surface shape, orientation, position, motion, and so forth, is processed by a human brain (Cumming, Johnston, and Parker, 1993; Landy et al., 1995; Jacobs, 2002; Knill and Saunders, 2003). Researchers widely used a concept of an ideal observer to define the information content used to achieve a certain level of performance (Blake, Bülthoff, and Sheinberg, 1993; Buckley, Frisby, and Blake, 1996; Knill, 1998; Geisler, 2003). It was hypothesized that visual information is extracted and initially processed in little quasi-independent pieces called *cues.* The alternative to this modular approach is to analyze a visual scene in all its complexity and interactions (i.e., strong fusion). This does not seem plausible due the harsh requirements on processing resources and learning time (Clark and Yuille, 1990; Johnston, Cumming, and Parker, 1993; Landy et al., 1995; Nakayama and Shimojo, 1996).

In short, visual cues are simple features that are relatively easy to extract and that correlate with important characteristics of visual scenes. For example, shape and size gradients of texture elements can be considered as cues for surface orientation. The assumption that cues indicate a certain scene property independently significantly simplifies cue learning and processing. Consequently, cue learning can be accomplished in isolation based, for example, on the correlation of a given cue with other cues from the same or different modality (Ernst and Banks, 2002; Ivanchenko and Jacobs, 2004). Cue processing includes a simple mechanism for combining the evidence from independent cues to improve overall accuracy of estimation.

### Classical Cue Integration Framework

A classical cue integration framework describes how several cues are combined at a single location. Because cues are assumed to be independently generated by a visual scene (or, in other words, the noise present in those cues is independent), the optimal rule for cue integration is essentially linear. Numerous experiments demonstrated that human performance corresponds well to the results of optimal cue integration (Young, Landy, and Maloney, 1993; Knill and Saunders, 2003; Hillis et al., 2004). Similar results were found when cues were integrated across different modalities such as visual, audio, and tactile (Ernst, Banks, and Bülthoff, 2000; Ernst and Banks, 2002; Battaglia, Jacobs, and Aslin, 2003).

Mathematically, one can express cue independence as the product of conditional probabilities. In addition, one can rewrite a probability of a scene property $S$ conditioned on cue $I1$ and $I2$ according to a Bayesian rule. This allows inferences to be made about a scene property (a cause) from observed cues (the effect):

$$P(S \mid I_1, I_2) = \frac{P(I_1, I_2 \mid S)P(S)}{P(I_1, I_2)} \alpha P(I_1 \mid S)P(I_2 \mid S)P(S) \qquad (5.1)$$

Here conditional probabilities on the right-hand side describe how cues $I1$, $I2$ are generated by a scene property $S$, and $P(S)$ specifies prior information about $S$. The denominator can be ignored because it is constant for a given scene. In the case of independent Gaussian noise, Equation (5.1) leads to a simple linear rule for cue integration (Cochran, 1937):

$$\hat{S} = \omega_1 \hat{S}_1 + \omega_2 \hat{S}_2 \qquad (5.2)$$

Here, $\hat{S}_1, \hat{S}_2$ are the estimates of a scene property from individual cues; $\hat{S}$ is a combined cue estimate; and $\omega_1, \omega_2$ are cue weights. It can be shown that weights are inversely proportional to cue uncertainty (i.e., variance of the noise) and that the uncertainty of a combined estimate is smaller than the uncertainty of any of its constituents. Cue uncertainty and, consequently, a cue weight change across visual conditions, and one of the important functions of the visual system is to adjust cue weights accordingly. When weights are known, Equation (5.2) provides a simple way to integrate estimates from individual cues to improve the overall accuracy of estimation.

## Cue Conflict

Because independence of visual cues is an approximation, the performance can be further improved if the perceptual system compensates for deviations from a linear rule. One of such nonlinearities, called a *cue conflict*, arises when cue estimates are too discrepant to be meaningfully combined. This discrepancy was initially explained by the presence of processing errors and outliers (Landy et al., 1995). To avoid the erroneous estimates, a weaker cue is down weighted or even dropped from cue combination. This makes cue integration robust to the presence of outliers.

Such heuristic treatment of a cue conflict was recently rationalized in a probabilistic Bayesian framework by associating a large cue conflict with incorrect prior assumptions (Knill, 2003, 2007). Prior assumptions (not to be mistaken with a prior probability of a scene property) are what make cues informative about a visual scene. For example, if one assumes that the world consists of circular objects, the aspect ratio of a projected circular outline would be a good cue for object orientation. However, if objects in the world have ellipsoidal outlines (less constrained prior), the strength of the aspect ratio cue would be greatly reduced. A key idea in this example is the assumption that the world consists of a mixture of circular and ellipsoidal objects.

Under different viewing conditions, each class of objects may have different prior probabilities of occurrence. This can be modeled with a mixture of priors that specifies probabilities for each object class. Then, preferential attenuation of a cue during a conflict would be based on the evidence for each prior model. Mathematically, the mixture of priors for a cue likelihood describes how the cue is generated by different classes of objects:

$$P(I|S) = \pi_1 * P(I|S, M_1) + \pi_2 * P(I|S, M_2) \cdots \qquad (5.3)$$

Here, the probabilities associated with each prior assumption $M_1, M_2$ are weighted by $\pi_1, \pi_2$ to form a full likelihood model for a cue $I$. The weights reflect the degree to which each prior assumption applies to a current environment. A large conflict provides evidence that an object was drawn from an ensemble with a less-constrained prior. According to Equation (5.3), a cue attenuation effect happens when a less-constrained prior dominates a mixture. Because a less-constrained prior entails a less informative cue, a cue weight goes down during a cue conflict.

### Example of Robust Cue Integration

In order to demonstrate some basic concepts of cue integration, I designed a simple visual example, shown in Figure 5.1. Here the shape of visual surfaces is represented by four cues: texture, shading, occlusion, and contour. Note that not all the cues are available at the same time. Particularly, at a small slant (top row) only texture and shading cues are present, while at a larger slant (bottom row) all four cues are present. The cues in the left column are consistent with the surface shape, and in order to demonstrate a cue conflict, the shading cue in the right column was made inconsistent with surface shape.

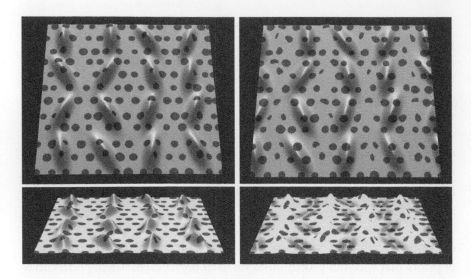

FIGURE 5.1 Robust cue integration for surface shape. In the left column, cues are consistent. In the right column, a shading cue is inconsistent with surface shape. In the top row, the surfaces are slanted at 30° from vertical, and in the bottom row the surfaces are slanted at 70°. A cue conflict is evident only at a large slant.

At a small slant, a shading cue for surface shape is the strongest one. According to a cue integration framework, texture and shading cues are combined, and the shading cue dominates the combination. When cues are inconsistent, a texture cue that has a smaller weight is dropped out of cue combination (upper right image). The perception of shape in this case is based solely on shading and is completely realistic in spite of the conflict. This demonstrates the robustness of cue integration in the presence of outliers (even though the outlier is a correct texture cue).

To demonstrate how cue strength varies across visual conditions, consider what happens when the surface slant is large (bottom row). At a large slant, two additional cues for shape became available—namely, a contour cue and an occlusion cue. Now the contour cue is the strongest one, and the remaining cues contribute their estimates to a lesser degree. In the case when cues are consistent (bottom left), the perception of shape remains intact. In the case when cues are inconsistent (bottom right), a relatively weaker shading cue is dropped out of the cue combination. This happens because a shading cue becomes an outlier due to its inconsistency with a stronger contour cue. Consequently, at a large slant, shading is no longer associated with surface shape and is perceived as a color of the surface (bottom right). Thus, the strength of visual cues changes with visual conditions and defines how visual cues are combined or ignored.

## Cue Promotion

Another potential nonlinearity of cue integration happens when a cue probability becomes multimodal or several cues depend on a common parameter (Landy et al., 1995; van Ee, Adams, and Mamassian, 2003; Adams and Mamassian, 2004). In the case when cue likelihood is multimodal, cues can interact with each other for the purpose of disambiguation: one cue can help another to select a particular peak in its likelihood function. When cues depend on a common parameter, interaction happens for the purpose of cue promotion. One goal of cue promotion is to eliminate mutual dependencies of cues on a common parameter; another goal is to expresses cues on a common scale so that they can be meaningfully combined. For example, both binocular disparity and kinetic depth effect cues scale with a visual distance (a common parameter) but in essentially different ways. Only at one viewing distance, they produce a consistent depth image, which can be used to solve for a viewing distance and to restore cue independence.

Although a particular mechanism of cue promotion remains largely unknown, it does not seem to represent a significant challenge on the algorithmic level. In fact, Landy et al. (1995) suggested several methods for how to solve for a common parameter of depth cues. However, if a cue for a certain scene property also depends on surface shape that changes as a function of spatial location, solving jointly for the cue and shape may be nontrivial (Ivanchenko, 2006). A more computationally plausible solution is to estimate surface shape independently from other scene properties. In a sense, this is similar to independent estimations of cues. A further discussion of this issue is given in the next section.

To summarize, the cue integration framework makes several major assumptions about information extraction and processing, such as the use of simple visual features (cues) and cue independence. On the one hand, these assumptions are quite general and simplify computation and learning. On the other hand, the assumptions are only approximately true. The visual system seems to compensate for simple dependencies between cues using cue promotion and disambiguation. To further improve performance, the visual system adjusts the weights on cues and priors depending on viewing conditions. Overall, a cue integration framework makes a basis for simple, robust, and statistically optimal estimation of scene properties when cues are considered at a single spatial location.

## CUE PROPAGATION IN THE CASE OF NOISY CUES

Cue propagation extends a cue integration framework by allowing the combining of visual cues at different spatial locations. The benefits of such integration are straightforward only if a visual surface has similar properties in some neighborhood; hence, we consider only planar or slowly curving surfaces. In those cases, highly informative but sparse cues can substitute for less informative neighbors due to the redundancy of visual information.

A mechanism for propagation of surface information was described by James Coughlan in detail in Chapter 3 of this book, "Mechanisms for Propagating Surface Information in 3D Reconstruction." This mechanism is not only tractable in terms of computation and learning times but also seems to be physiologically plausible. It also allows information to be propagated along a surface without knowing surface position and orientation in advance. Merely specifying a prior constraint such as smoothness of some surface parameter is enough to produce a consistent and dense surface representation. Such constraints specify how information is interpolated during propagation.

Here I look at this mechanism from a viewpoint of cue integration theory and make some parallels between computational formulation and cue probabilities. I focus on the case when visual information is propagated into a surface area with noisy and unreliable cues and show that this situation is analogous to a cue conflict. The main goals are to analyze the conditions for cue propagation and compare the role of different prior constraints in surface reconstruction. As an example, I use three-dimensional (3D) surface reconstruction from binocular disparity that is similar to that described by Coughlan in Chapter 3 of this book. Finding binocular correspondences remains a hot topic in computer vision (Sun, Shum, and Zheng, 2002; Zhang and Seitz, 2005; Seitz et al., 2006). Some of the challenges are scarcity of strong visual cues and presence of multiple matching hypotheses. Though the latter characteristic is different from the assumption of the classical cue integration framework (where a unimodal Gaussian distribution represents a single hypothesis), the same principles apply. For example, as the strength of a cue declines due to the changes in viewing conditions, the role of prior information correspondingly increases.

The model of information propagation described by Coughlan in Chapter 3 of this book is based on probabilistic formulation of a surface as a Markov random field (MRF). MRFs are widely used in computer vision to express the joint probability of a grid of random variables with local interaction (Weiss, 1997; Weiss and Freeman, 2001). Similar to cue integration, MRFs combine evidence from image cues and prior cue probabilities. Unlike cue integration, MRFs specify a constraint on neighboring variables to represent local interactions.

There are several methods for performing inference with MRFs that maximize either marginal or maximum *a posteriori* (MAP) probabilities. These methods were proven to perform optimally in the absence of loops in MRF but were also shown to perform well in loopy cases (Cochran, 1937; Felzenszwalb and Huttenlocher, 2006). One such method, belief propagation (BP), is naturally suited for modeling information propagation as was suggested in Coughlan in Chapter 3 of this book. This is because BP explicitly describes how neighboring variables exchange messages related to hypotheses about local surface properties.

For the purposes of the current discussion, it is sufficient to say that during BP, each variable in MRF sends a message to its neighbors, and this process iterates throughout the grid until messages converge. The content of these messages reflects a best sender's guess about the receiver's likelihood based on all locally available evidence including a prior. If we associate a discrete likelihood with a set of hypotheses about a visual property (e.g., disparity in the image), then we can say that through message updates, BP dynamically reevaluates all possible hypotheses based on the compromise between local evidence and a prior constraint.

Note that BP has no explicit mechanism for propagation of strong cues (i.e., those whose likelihoods have strong peaks) into the locations that contain weak cues (i.e., those whose likelihoods have no strong peaks). Nevertheless, such propagation does happen in the MRF framework. Moreover, the propagation can be facilitated if the regions where cues propagate contain variables with flat likelihood. Such a case was described by Coughlan in Chapter 3 of this book, where disparity information was propagated from a textured image region into a textureless one. Note that the textureless region has no cues for disparity; thus, corresponding likelihoods are flat. The propagation happens because MAP to which an algorithm converges is a product of likelihoods and smoothness priors, and the latter are higher for variables with similar likelihoods. Thus, initially flat likelihoods inside of the textureless region acquire values similar to the likelihoods at the region periphery. Importantly, in this case, the region with flat likelihoods has no evidence that is contrary to the one being propagated into the region.

Here we consider a case when cues are present throughout the image but weak or noisy at some of its regions (as happens in the areas with low contrast texture). Consequently, cue likelihoods are not flat but rather have a few weakly pronounced peaks that correspond to multiple matching hypotheses. Some peaks happen as a result of random noise in one or both images of a stereo pair. Though these noisy likelihoods look similar to flat ones, propagation is limited. As computer simulations show, variables at the regions with noisy likelihoods converge to MAP probabilities that show little or no influence from neighboring regions with strong likelihoods. Thus, propagation is limited in the case when likelihoods are not completely flat.

Algorithmically, this can be explained by the fact that when two variable likelihoods inside of a noisy region coincidentally express similar hypotheses, these hypotheses are reinforced by a smoothness prior. Variables with such reinforced likelihoods become a major obstruction to propagation. This is because the hypotheses expressed in their likelihood often do not support the ones propagated into the region (and vice versa). More research is required in order to better understand factors for information propagation in MRFs. Here I suggest some solutions for improving propagation in the areas with noisy cues.

A straightforward solution is to discard the regions with noisy or weak cues from consideration. This is a common practice for a disparity validation procedure in correlation-based stereo. The obvious disadvantage of this procedure is holes in a disparity map. Another solution is to use multiscale MRFs where short-distance propagation on a rough scale corresponds to long-distance propagation on a fine scale. However, a more principled approach is to detect weak likelihoods and flatten them out. A tentative definition of a discrete weak likelihood can be based on a comparison of the magnitude of its peak with an average value expected for a single level. Likelihood is weak if its peaks are not much larger than the expected mean value. Note that detecting weak likelihoods is similar to finding weak cues during a cue conflict. Moreover, a flattening process is similar to ignoring a discrepant cue during robust cue integration. This is because a flat likelihood carries no hypothesis and thus cannot be considered as a cue.

It is interesting to note that although a likelihood flattening method was derived solely from practical consideration, there is theoretical justification for it. It is based on the probabilistic theory of a cue conflict discussed in the first section. Mathematically, this theory describes a mixture of prior assumptions that make cues informative. For the problem of finding binocular correspondence, one can model images as a mixture of at least two types of objects. One contains strong edges (e.g., outlines of objects and high contrast texture), and the other includes image regions with low-contrast intensities (e.g., uniform color of object surfaces and low contrast texture). For the first class of objects, we can use very informative edge cues, and for the second class of objects, the cues are less informative. According to the mixture of priors approach, we can express a full cue likelihood as

$$P(I|S) = \pi_{edge}P(I|S, M_{edge}) + \pi_{intensity}P(I|S, M_{intensity})$$  (5.4)

Here the first component of the likelihood is due to informative edge cues, and the second component comes from less informative intensity cues. The strength of the edge can be used to indicate the weight to which each prior model applies. Then a formula for likelihood (Coughlan, 2011, this volume ) that expresses a matching error, $m(D)$ at a certain image location and disparity $D$,

$$P(m(D)|D) = \frac{1}{Z}e^{-\mu m(D)}$$  (5.5)

can be rewritten as

$$P(m(D)|D) = \frac{1}{Z'}\left[\pi_{edge}e^{-\mu_1 m(D)} + \pi_{intensity}e^{-\mu_2 m(D)}\right]$$  (5.6)

Where $\mu 2 < \mu 1$ to reflect the fact that an intensity cue is less informative than the edge cue. In the marginal case of $\mu_2 = 0, \pi_{intensity} = 1$, and $\pi_{edge} = 0$, we obtain the above-described flattening method for likelihoods. Note that in order to detect cues with weak likelihoods, one can directly analyze likelihood peaks, which is more beneficial than measuring edge strength. The former has an advantage of finding cases when the reason for weak likelihood is not only a low contrast texture but also a highly regular texture (i.e., the one that produces multiple matching hypotheses).

As computer simulations show, flattening weak likelihoods greatly improves propagation of disparity information into regions with weak cues. Such a method is especially applicable in stereo algorithms, because they often rely on some smoothness constraint that justifies cue propagation

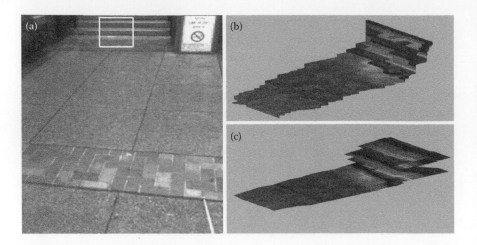

FIGURE 5.2 Surface reconstruction from binocular disparity. Left (a): region of one of the images from a stereo pair; area for reconstruction is depicted with a white rectangle. Top right (b): surface reconstruction based on disparity smoothness prior. Bottom right(c): surface reconstruction based on elevation smoothness prior.

and provides a reasonable way to interpolate information along the surface. The only image regions where smoothness constraint is usually not enforced correspond to borders of objects; but those areas typically have high-contrast pixels with strong cues and thus are not affected by propagation. Thus, a cue conflict theory seems to be applicable to at least one computer vision problem where it facilitates information propagation and the creation of dense stereo maps from sparse cues.

Figure 5.2 shows 3D reconstruction of a surface region based on the above described flattening method. Note that reconstructed surfaces are dense and have no holes that are typical for a correlation-based stereo. It was possible to run a BP algorithm in near real time (less than 200 ms on GPU) with VGA image resolution and 32 disparity levels (Ivanchenko, Shen, and Coughlan, 2009).

We reconstructed a distant image region on purpose because the strength of a disparity cue decreases with the viewing distance. This makes distant regions good candidates for observing weak cues and for analyzing the role of a smoothness constraint that increases when a cue strength decreases. Note that this constraint describes properties of visual surfaces in the world as opposed to a mixture of priors that describes properties of the images.

In 3D reconstruction, we used two different smoothness constraints: one enforcing smooth disparity and one enforcing smooth elevation of visual surfaces. These constraints provide a way for automatic interpolation of information along the surface and also bias surface reconstruction to produce either fronto-parallel surface patches (disparity prior) or patches that are parallel to the ground plane (elevation prior). As can be seen in Figure 5.2, the reconstructed surfaces of a staircase have different shapes depending on a corresponding smoothness assumption. Two reconstructions look similar only at the locations where cues are strong (i.e., have strong edges).

Note that while a flattening method helps to obtain a dense surface representation, the form of representation in the areas with weak cues is affected by a smoothness constraint. Which reconstruction and corresponding smoothness constraint is better? There is no single answer to this question. The factors that influence a choice of a constraint include the structure of the environment, task requirements, and the compactness of a resulting representation (e.g., a number of disparity levels).

To summarize, this section analyzed the information propagation method described by Coughlan in Chapter 3 of this book from a cue integration perspective. Specifically, we considered the case when strong cues are propagated across flat or slowly curving surfaces into the areas with weak or noisy cues. To guarantee that propagation is efficient, it was suggested that cue likelihoods that have only weak peaks should be flattened completely. This was justified under a probabilistic approach that considers a mixture of prior assumptions for each cue. For the purpose of finding a dense disparity map, stereo images were modeled as a mixture of two classes of objects: one that has strong edges and one with low-contrast intensities. Note that the first class of object is typically sparse in natural images and has strong cues; the second class of object is less sparse but has weaker cues. To define a prior probability of each class in a certain image location, one can use edge strength or directly analyze cue likelihood. While mixture of priors describes image properties, there is another class of priors that describe surfaces. This section analyzed the effect of two such priors (smoothness of elevation and disparity) on 3D reconstruction.

## CUE INTERPRETATION FOR NONFLAT SURFACES

So far we described how cues are integrated at a single location and how they can be propagated to different locations along a flat or slowly curving surface. In the general case of nonflat surfaces, we have to deal with the fact that visual cues for some surface property may also depend on surface shape. To give an example, consider a texture foreshortening cue for slant. In the case of nonflat surfaces, this cue depends on both surface slant and shape. Thus, in order to interpret this cue in terms of slant, the effect of shape should be disentangled from the effect of slant. Furthermore, if several cues are present at the same location on a nonflat surface, they are no longer independent due to the explaining away phenomenon. For instance, different cues for slant can mutually depend on a surface shape. This violates the assumptions of linear cue combination theory and potentially can make cue integration a difficult problem.

One of the solutions to this problem is given by a modular approach. Similar to the case of cue integration when each cue is processed independently by a separate module, we can assign separate modules for estimation of surface shape and slant. Then, in a slant module, for example, the information about shape would be represented with a fixed prior. A limited interaction would take place after estimation is finished, and modules exchange information in order to update their priors.

Another observation that supports a modular approach is that cues for shape can be quite complicated and specific. Consider, for example, a shading cue for shape. This cue depends on the vector of light source direction, light spectrum, surface reflectance properties, and so forth. It is difficult to imagine a mechanism for interaction between this specific cue and the cues for surface slant. On the other hand, if shape is represented in a cue independent form (such as surface curvature) at the output of a shape-processing module, it would be much easier to incorporate such a representation into the interpretation of slant cues. Thus, a simplicity principle suggests that surface shape should be estimated in isolation and used for the proper interpretation and integration of surface cues. Moreover, according to principles described in the second section, shape information can be used to propagate sparse cues along nonflat surfaces. In other words, estimation of shape can help to automatically interpolate surface cues in locations where they are sparse.

A perceptual mechanism for shape processing is yet to be studied, but we aim to establish whether or not the human visual system can use shape information to interpret surface cues properly. To investigate this issue, we conducted two psychophysical experiments that are described in detail in Ivanchenko (2006). These experiments used a slant discrimination task to analyze the role of shape cues on subject performance. The

FIGURE 5.3 Cue conditions on a slant discrimination task. Texture (left) was generated by a reaction-diffusion process and was statistically uniform. Shading (middle) was a mix of diffuse and ambient components with a point light source placed above the surface. A combined cue condition (right) had both cues. The surface shape was a mixture of two-dimensional Gaussians that were aligned on a grid and had a roughly vertical orientation.

visual cues were shading and texture. Note that these cues are affected by both slant and shape of a surface. In other words, the cues for slant mutually depend on the surface shape and vice versa. Stimulus shape was represented with corrugated surfaces that were planar on a large scale and had a pattern of roughly vertical ridges on a small scale. Surfaces were rendered in three conditions: texture, shading, and combined cues (see Figure 5.3).

Note that stimuli in shading condition have very weak cues for slant, and there is a bias toward a fronto-parallel interpretation. To reduce this bias, we used a slant discrimination task that included trials with two sequential presentations of a surface in the same condition. In each trial, subjects had to pick a surface that had a greater slant. The subjects' performance was summarized with a psychometric function.

In order to understand how the visual system processes slant information, we compared subject performance with the performance that was expected from a linear cue combination. As we show below, a linear model assumes that shape is not estimated when estimating slant. Alternatively, the visual system can use the information about surface shape for slant estimation. To measure performance on a discrimination task, we calculated a threshold of a psychometric function. Assuming Gaussian noise, the threshold is proportional to the standard deviation of the noise. According to a linear rule, the thresholds in texture ($\sigma_{tex}$) and shading

($\sigma_{shad}$) conditions can be used to calculate the expected linear threshold ($\sigma_{linear}$) in a combined condition:

$$\frac{1}{\sigma^2_{linear}} = \frac{1}{\sigma^2_{tex}} + \frac{1}{\sigma^2_{shad}} \tag{5.7}$$

Note that such a rule is applicable only if the visual system ignores the shape (e.g., a planar approximation is used instead). If the visual system uses shape estimates to inform the estimation of slant, the linear rule is suboptimal. This is because the linear rule applied to slant discrimination describes the integration of two slant cues but ignores their interaction with shape cues. Optimally, in combined cue condition, the overall accuracy of shape estimation increases and can further improve slant estimation given that these processes interact. We found that a mean threshold in combined cue condition was significantly lower than the expected linear threshold ($t$-test, $p = 0.046$). Thus, subjects performed much better than was expected from a linear rule indicating that shape and slant estimation processes interact.

To get additional evidence about the role of shape cues in slant discrimination, we conducted a second experiment where we decoupled texture and shading cues. Specifically, we made the latter one uninformative for interpreting cues for slant. This was achieved by a horizontal shift of texture projection in the image. Despite such drastic manipulation, a cue conflict was hardly noticeable, and texture was still perceived attached to a visual surface depicted by shading. After this manipulation, the threshold in combined cue condition significantly increased compared to the one measured in the first experiment ($t$-test, $p = 0.015$). Because our stimulus manipulation created a cue conflict for shape cues but not for slant cues, we concluded that accurate shape information was crucial for interpretation of slant cues. Overall, the results of two psychophysical experiments suggest that using shape information improves subject performance on a slant discrimination task, and the absence of consistent shape information significantly deteriorates subject performance. We concluded that shape cues are used by the human visual system to properly interpret slant cues.

## CONCLUSIONS

In this chapter, we reviewed a mechanism for information propagation described by Coughlan in Chapter 3 of this book from a viewpoint of a cue integration theory. We found many similarities between a computational model of information propagation and underlying principles of cue integration. We also found that a cue conflict theory can be used to improve information propagation into areas with weak and noisy cues. Using this improvement, we obtained dense 3D surface reconstructions from a stereo pair with sparse strong cues. Importantly, the shape of 3D reconstruction depended on the form of a prior constraint that helped to interpolate information along flat or slowly curving surfaces. For nonflat surfaces, we showed that the human visual system uses shape information to increase the informativeness of visual cues for slant. This is consistent with the hypothesis that surface shape is estimated in isolation to facilitate interpretation of other cues and interpolate them in the location where cues are sparse.

## ACKNOWLEDGMENT

I would like to thank James Coughlan, David Knill, and Ender Tekin for helpful feedback on this manuscript.

# CHAPTER 6

# Symmetry, Shape, Surfaces, and Objects

Tadamasa Sawada

Yunfeng Li

Zygmunt Pizlo

## CONTENTS

## INTRODUCTION

Traditionally, the perception of three-dimensional (3D) scenes and objects was assumed to be the result of the reconstruction of 3D surfaces from available depth cues, such as binocular disparity, motion parallax, texture, and shading (Marr, 1982). In this approach, the role of figure–ground organization was kept to a minimum. Figure–ground organization refers to finding objects in the image, differentiating them from the background, and grouping contour elements into two-dimensional (2D) shapes representing the shapes of the 3D objects. According to Marr, solving figure–ground organization was not necessary. The depth cues could provide the observer with local estimates of surface orientation at many points. These local estimates were then integrated into the estimates of the visible parts of 3D shapes. Marr called this a 2.5D representation. If the objects in the scene are familiar, their entire 3D shapes can be identified (recognized) by matching the shapes in the observer's memory to the 2.5D sketch. If they are not familiar, the observer will have to view the object from more than one viewing direction and integrate the visible surfaces into the 3D shape representations.

Reconstruction and recognition of surfaces from depth cues might be possible, although there has been no computational model that could do it reliably. Human observers cannot do it reliably, either. These failures suggest that the brain may use another method to recover 3D shapes, a method that does not rely on depth cues as the only or even the main source of shape information. Note that there are two other sources of information available to the observer: the 2D retinal shapes produced by figure–ground organization and 3D shape priors representing the natural world statistics. Figure–ground organization was defined in the first paragraph and will be discussed next. The 3D shape priors include 3D symmetry and 3D compactness. These priors have been recently used by the present authors in their models of 3D shape perception (Pizlo, 2008; Sawada and Pizlo, 2008a; Li, 2009; Li, Pizlo, and Steinman, 2009; Sawada, 2010; Pizlo et al., 2010), and they will be discussed in the following section.

## FIGURE–GROUND ORGANIZATION

The concept of figure–ground organization was introduced by Gestalt psychologists in the first half of the 20th century (Wertheimer, 1923/1958; Koffka, 1935). Gestalt psychologists emphasized the fact that we do not perceive the retinal image. We see objects and scenes, and we can perceptually judge their shapes, sizes, and colors, but we cannot easily (if at all) judge their retinal images. The fact that we see shapes, sizes, and colors of objects veridically despite changes in the viewing conditions (viewing orientation, distance, and chromaticity of the illuminant) is called *shape, size*, and *color constancy*, respectively. Changes in the viewing conditions change the *retinal image*, but the *percept* remains constant. How are those constancies achieved? Gestalt psychologists pointed out that the percept can be explained by the operation of a simplicity principle. Specifically, according to Gestaltists, the percept is the simplest interpretation of the retinal image. Figures 6.1 and 6.2 provide illustrations of how a simplicity principle can explain the percept. Figure 6.1 can be interpreted as two overlapping rectangles or two touching hexagons. It is much easier to see the former interpretation because it is simpler. Clearly, two symmetric rectangles is a simpler interpretation than two asymmetric hexagons. Figure 6.2a can be interpreted as two lines forming an *X* or two *V*-shaped lines touching

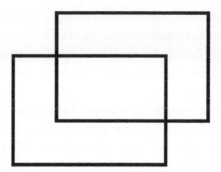

FIGURE 6.1   Two rectangles is a simpler interpretation than two concave hexagons.

at their vertices (the *V*s are perceived as either vertical or horizontal). The reader is more likely to see the former interpretation because it is simpler. In Figure 6.2b, the reader perceives two closed shapes touching each other. Note that Figure 6.2b was produced from Figure 6.2a by "closing" two horizontal *V*s. Closed 2D shapes on the retina are almost always produced by occluding contours of 3D shapes. Therefore, this interpretation is more likely. So, global closure is likely to change the local interpretation of an *X*. The fact that spatially global aspects, such as closure, affect spatially local percepts refutes Marr's criticism of Gestalt psychology. Marr (1982) claimed that all examples of perceptual grouping can be explained by recursive application of local operators of proximity, similarity, and smoothness. If Marr were

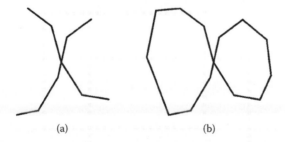

(a)                              (b)

FIGURE 6.2   Spatially global features (closure) affect spatially local decisions. (a) This figure is usually perceived as two smooth lines forming an X intersection. (b) This figure is usually perceived as two convex polygons touching each other at one point. Apparently, closing the curves changes the interpretation at the intersection, even though the image around the intersection has not changed.

right, than the perceptual "whole would be equal to the sum of its parts." Figure 6.2b provides a counterexample to Marr's claim. In general, forming perceptual shape cannot be explained by summing up the results of spatially local operators. The operators must be spatially global.

It is clear that the percept is more organized, richer, simpler, as well as more veridical than the retinal image. This fact demonstrates that the visual system adds information to the retinal image. Can this "configurality" effect be formalized? Look again at Figure 6.1. You see two "perfect" rectangles. They are perfect in the sense that the lines representing edges are perceived as straight and smooth. This is quite remarkable considering the fact that the output of the retina is more like that shown in Figure 6.3. Specifically, the retinal receptors provide a discrete sampling of the continuous retinal image, and the sampling is, in fact, irregular. The percept of continuous edges of the rectangles can be produced by interpolation and smoothing. Specifically, one can formulate a cost function with two components: the first evaluates the difference (error) between the perceptual interpretation and the data, and the second evaluates the complexity of the perceptual interpretation. The percept corresponds to the minimum of the cost function. Clearly, two rectangles with straight-line edges are a better (simpler) interpretation than a set of unrelated points, or than two rectangles with jagged sides. This example shows that the traditional, Fechnerian approach to perception is inadequate.

According to Fechner's (1860/1966) approach, the percept is a result of retinal measurements. It seems reasonable to think that such properties

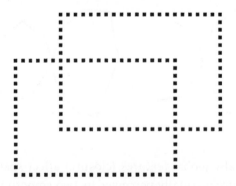

FIGURE 6.3 When the retinal image is like that in Figure 6.1, the input to the visual system is like that shown here. Despite the fact that the input is a set of discrete points, we see continuous rectangles when we look at Figure 6.1.

of the retinal image as light intensity or position could be measured by the visual system. It is less obvious how 3D shape can be measured on the retina. When we view a Necker cube monocularly, we see a cube. This percept cannot be a *direct* result of retinal measurement, because this 2D retinal image is actually consistent with infinitely many 3D interpretations. The only way to explain the percept is to assume that the visual system "selects" the simplest 3D interpretation. Surely, a cube, an object with multiple symmetries, is the simplest 3D interpretation of any of the 2D images of the cube (except for several degenerate views). Note also that a cube is a maximally compact hexahedron. That is, a cube is a polyhedron with six faces whose volume is maximal for a given surface area. Recall that both symmetry and compactness figured prominently in the writings of the Gestalt psychologists (Koffka, 1935).

We now know that it is more adequate to "view" 3D shape perception as an inverse problem, rather than as a result of retinal measurements (Pizlo, 2001, 2008). The forward (direct) problem is represented by the retinal image formation. The inverse problem refers to the perceptual inference of the 3D shape from the 2D image. This inverse problem is ill posed and ill conditioned, and the standard (perhaps even the only) way to solve it is to impose *a priori* constraints on the family of possible 3D interpretations (Poggio, Torre, and Koch, 1985). In effect, the constraints make up for the information lost during the 3D to 2D projection. The resulting perceptual interpretation is likely to be veridical if the constraints represent regularities inherent in our natural world (natural world statistics). If the constraints are deterministic, one should use regularization methods. If the constraints are probabilistic, one should use Bayesian methods. Deterministic methods are closer to the concept of a Gestalt simplicity principle, whereas probabilistic methods are closer to the concept of a likelihood principle, often favored by empiricists. Interestingly enough, these two methods are mathematically equivalent, as long as simplicity is formulated as an economy of description. This becomes quite clear in the context of a minimum description length approach (Leclerc, 1989; Mumford, 1996). The close relation between simplicity and likelihood was anticipated by Mach (1906/1959) and discussed more recently by Pomerantz and Kubovy (1986). Treating visual perception as an inverse problem is currently quite universal, especially among computational modelers. This approach has been used to model binocular vision, contour, motion, color, surface, and shape perception (for a review, see Pizlo, 2001).

It follows that if one is interested in studying the role of *a priori* constraints in visual perception, one should use distal stimuli that are as different from their retinal images as possible. This naturally implies that the distal stimuli should be 3D objects or images of 3D objects. In such cases, the visual system will have to solve a difficult inverse problem, and the role of constraints can be easily observed. If 2D images are distal stimuli, as in the case of examples in Figures 6.1, 6.2, and 6.3, *a priori* constraints may or may not be involved depending on whether the visual system "tries" to solve an inverse problem. As a result, the experimenter may never know which, if any, constraints are needed to explain a given percept. Furthermore, veridicality and constancy of the percept are poorly, if at all, defined with such stimuli. Also, if effective constraints are not applied to a given stimulus, then the inverse problem remains ill posed and ill conditioned. This means that the subjects' responses are likely to vary substantially from condition to condition. In such a case, no single model will be able to explain the results. This may be the reason for why the demonstrations involving 2D abstract line drawings (like those in Kanizsa's 1979 book) are so difficult to explain by any single theory. Figure–ground organization and amodal completion are hard to predict when the distal stimulus is 2D. We strongly believe that figure–ground organization should be studied with 3D objects. Ambiguities rarely arise with images of natural 3D scenes, because the visual system can apply effective constraints representing natural world statistics.

We want to point out that perceptual mechanisms underlying figure–ground organization are still largely unknown. But we hope that our 3D shape recovery model will make it easier to specify the requirements for the output of figure–ground organization. Once the required output of this process is known, it should be easy to study it. The next section describes our new model of 3D shape recovery. A discussion of its implications is presented in the last section.

## USING SIMPLICITY PRINCIPLE TO RECOVER 3D SHAPES

We begin by observing that 3D shape is special because it is a complex perceptual characteristic, and it contains a number of regularities. Consider first the role of complexity. Shapes of different objects tend to be quite different. They are different enough so that differently shaped objects never produce identically shaped retinal images. For example, a chair and a car will never produce identically shaped 2D retinal images, regardless of their viewing directions, and similarly with a teapot, book, apple, and so forth. This implies that each 3D shape can be uniquely recognized from any of its 2D images. This fact is critical for achieving shape constancy. More precisely, if the observer is familiar with the 3D shape, whose image is on his or her retina, this shape can be unambiguously identified. What about the cases when the 3D shape is unfamiliar? Can an unfamiliar shape be perceived veridically? The answer is yes, as long as the observer's visual system can apply effective shape constraints to perform 3D shape recovery. Three-dimensional symmetry is one of the most effective constraints. It reduces the ambiguity of the shape recovery problem substantially. The role of 3D symmetry will be explained next.

Figure 6.4a consists of 13 points, images of vertices of a 3D symmetric polyhedron shown in Figure 6.4b. The polyhedron in Figure 6.4b is perceived as a 3D shape; the points in Figure 6.4a are not. The points in Figure 6.4a are perceived as lying on the plane of the page. Formally, when a single 2D image of $N$ unrelated points is given, the 3D reconstruction problem involves $N$ unknown parameters, the depth values of all points. The problem is so underconstrained that the visual system does not even "attempt" to recover a 3D structure. As a result, we see 2D, not 3D, points. When points form a symmetric configuration in the 3D space, the reconstruction problem involves only one (rather than $N$) unknown parameter—namely, the 3D aspect ratio of the configuration (Vetter and Poggio, 1994; Li, Pizlo, and Steinman, 2009; Sawada, 2010). Note that the visual system must organize the retinal image first and (1) determine that the image could have been produced by a 3D symmetric shape and (2)

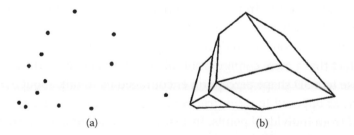

(a)                                           (b)

FIGURE 6.4  Static depth effect. The three-dimensional (3D) percept produced by looking at (b) results from the operation of 3D symmetry constraint when it is applied to an organized retinal image. The percept is not 3D when the retinal image consists of a set of unrelated points, like that in (a). (Reprinted with permission from Pizlo et al., 2010).

identify which pairs of image features are actually symmetric in the 3D interpretation. These two steps can be accomplished when the 2D retinal image is organized into one or more 2D shapes. Clearly, the visual system does this in the case of Figure 6.4b, but not Figure 6.4a. The only difference between the two images is that the individual points are connected by edges in Figure 6.4b. Obviously, the edges are not drawn arbitrarily. The edges in the 2D image correspond to edges of a polyhedron in the 3D interpretation. The difference between Figure 6.4a and Figure 6.4b is striking: the 3D percept in the case of Figure 6.4b is automatic and reliable. In fact, the reader cannot avoid seeing a 3D object no matter how hard he or she tries. Just the opposite is true with Figure 6.4a. It is impossible to see the points as forming a 3D configuration. Adding "meaningful" edges to 2D points is at least as powerful as adding motion in the case of kinetic depth effect (Wallach and O'Connell, 1953). We might, therefore, call the effect presented in Figure 6.4, a static depth effect (SDE).

How does symmetry reduce the uncertainty in 3D shape recovery? In other words, what is the nature of a symmetry constraint? A 3D reflection cannot be undone, in a general case, by a 3D rigid rotation, because the determinant of a reflection matrix is "−1," whereas the determinant of a rotation matrix is "+1." However, a 3D reflection is equivalent to (can be undone by) a 3D rotation when a 3D object is mirror symmetric. It follows that, given a 2D image of a 3D symmetric object, one can produce a virtual 2D image of the same object, by computing a mirror image of the 2D real image. This is what Vetter and Poggio (1994) did. A 2D virtual image is a valid image of the same 3D object, only if the object is mirror symmetric. So, a 3D shape recovery from a single 2D image turns out to be equivalent to a 3D shape recovery from two 2D images, when the 3D shape is mirror symmetric. It had been known that two 2D orthographic images determine the 3D interpretation with only one free parameter (Huang and Lee, 1989).

The fact that a mere configuration of points and edges in the 2D image gives rise to a 3D shape perception encouraged us to talk about 3D shape *recovery*, as opposed to *reconstruction*. The 3D shape is not actually *constructed* from individual points. Instead, it is produced (recovered) in one computational step. The one-parameter family of 3D symmetric interpretations consistent with a single 2D orthographic image is illustrated in Figure 6.5. The 2D image (line drawing) is shown on top. The eight shaded objects are all mirror symmetric in 3D, and they differ from one another with respect to their aspect ratio. The objects on the left-hand side are wide, and the objects on the right-hand side are tall. There are infinitely

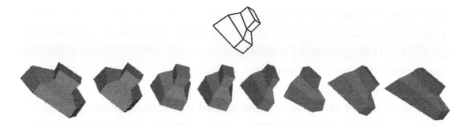

FIGURE 6.5 The shaded three-dimensional (3D) objects illustrate the one-parameter family of 3D shapes that are possible 3D symmetric interpretations of the line drawing shown on top.

many such 3D symmetric shapes. Their aspect ratio ranges from infinitely large to zero. Each of these objects can produce the 2D image shown on top. How does the visual system decide among all these possible interpretations? Which one of these infinitely many 3D shapes is most likely to become the observer's percept? We showed that the visual system decides about the aspect ratio of the 3D shape by using several additional constraints: maximal 3D compactness, minimum surface area, and maximal planarity of contours (Sawada and Pizlo, 2008a; Li, Pizlo, and Steinman, 2009; Pizlo et al., 2010). These three constraints actually serve the purpose of providing an interpretation (3D recovered shape) that does not violate the nonaccidental (generic) viewpoint assumption. Specifically, the recovered 3D shape is such that the projected image would not change substantially under small or moderate changes in the viewing direction of the recovered 3D shape. In other words, the 3D recovered shape is most likely. This means that it is also the simplest (Pizlo, 2008).

Several demonstrations illustrating the quality of 3D shape recovery from a single 2D image can be viewed at http://www1.psych.purdue .edu/~sawada/minireview/. In Demo 1, the shapes are abstract, randomly generated polyhedra. In Demo 2, images of 3D models of real objects, as well as real images of natural objects are used. In Demo 3, images of human bodies are used. In all these demos, the 2D contours (and skeletons) were drawn by hand and "given" to the model. First, consider the images of randomly generated polyhedra, like those in Figure 6.4b (go to Demo 1). The 2D image contained only the relevant contours—that is, the visible contours of the polyhedron. The model had information about where the contours are in the 2D image and which points and which contours are symmetric in 3D. Obviously, features that are symmetric in 3D space are not symmetric in the 2D image. Instead, they are skew-symmetric (Sawada and Pizlo, 2008a, 2008b). There is an invariant of the 3D to 2D projection of a mirror-symmetric object. Specifically, the lines connecting the 2D points that are images of pairs of 3D symmetric points are all parallel to one another in any 2D orthographic image, and they intersect at a single point (called a *vanishing point*) in any 2D perspective image. This invariant can be used to determine automatically the pairs of 2D points that are images of 3D symmetric points. Finally, the model had information about which contours are planar in 3D. Planarity constraint is critical in recovering the back, invisible part of the 3D shape (Li, Pizlo, and Steinman, 2009). We want to point out, again, that the information about where in the image the contours are and which contours are symmetric and which are planar in 3D corresponds to establishing figure–ground organization in the 2D image. Our model does not solve the figure–ground organization (yet). The model assumes that figure–ground organization has already been established. It is clear that the model recovered these polyhedral shapes very well. The recovered 3D shapes are accurate for most shapes and most viewing directions. That is, the model achieves shape constancy. Note that the model can recover the entire 3D shape, not only its front, visible surfaces. The 3D shapes recovered by the model were similar to the 3D shapes recovered by human observers (Li, Pizlo, and Steinman, 2009).

Next, consider examples with 3D real objects (Demo 2). One of us marked the contours in the image by hand. Note that the marked contours were not necessarily straight-line segments. The contours were curved, and they contained "noise" due to imperfect drawing. The reader realizes that marking contours of objects in the image is easy for a human observer; it seems effortless and natural. However, we still do not know how the visual system does it. After the contours were drawn, the pairs of contours that are symmetric in 3D and the contours that are planar were labeled by hand. Because objects like a bird or a spider do not have much volume and the volume is not clearly specified by the contours, the maximum compactness and minimum surface area constraints were applied to the convex hull of the 3D contours. The convex hull of a set of points is the smallest 3D region containing all the points in its interior or on its boundary. Again, it can be seen that the 3D shapes were recovered by the model accurately. Furthermore, the recovered 3D shapes are essentially identical to the shapes perceived by the observer. Our 3D shape recovery model seems to capture all the essential characteristics of human visual perception of 3D shapes.

Finally, consider the case of images of models of human bodies. Note that two of the three human models are asymmetric. The skeletons of humans were marked by hand, and the symmetric parts of skeletons were manually identified, as in the previous demos. Marking skeletons was no more difficult than marking contours in synthetic images. The most interesting aspect of this demo is the recovery of asymmetric human bodies. The human body is symmetric in principle, but due to changes in the position and orientation of limbs, the body is not symmetric in 3D. Interestingly, our model is able to recover such 3D asymmetric shapes. The model does it by first correcting the 2D image (in the least squares sense) so that the image is an image of a symmetric object. Then, the model recovers the 3D symmetric shape and uncorrects the 3D shape (again, in the least squares sense), so that the 3D shape is consistent with the original 2D image (Sawada and Pizlo, 2008a; Sawada, 2010). The recovered 3D shape is not very different from the true 3D shape and from the perceived 3D shape.

## SUMMARY AND DISCUSSION

We described the role of 3D mirror symmetry in the perception of 3D shapes. Specifically, we considered the classical problem of recovering 3D shapes from a single 2D image. Our computational model can recover the 3D structure (shape) of curves and contours whose 2D image is identified and described. Finding contours in a 2D image is conventionally referred to as figure–ground organization. The 3D shape is recovered quite well, and the model's recovery is correlated with the recovery produced by human observers. Once the 3D contours are recovered, the 3D surfaces can be recovered by interpolating the 3D contours (like wrapping the 3D surfaces around the 3D contours) (Pizlo, 2008; Barrow and Tenenbaum, 1981). It follows that our model is quite different from the conventional, Marr-like approach. In the conventional approach, 3D visible surfaces (2.5D sketch) were reconstructed first. Next, the "2.5D shape" was produced by changing the reference frame from viewer centered to object centered. But does a 2.5D representation actually have shape? Three-dimensional shape refers to spatially global geometrical characteristics of an object that are invariant under rigid motion and size scaling. However, 3D surfaces are usually described by spatially local features such as bumps, ridges, and troughs. Surely, specifying the 3D geometry of such surface features does say something about shape, but spatially local features are not shapes. Spatially local features have to be related (connected) to one another to form a "whole." Three-dimensional symmetry is arguably the best, perhaps even the only way to provide such a relational structure and, thus, to represent shapes of 3D objects. But note that visible surfaces of a 3D symmetric shape are never symmetric (except for degenerate views). If symmetry is necessary for shape, then arbitrary surfaces may not have shape. At least, they may not have shape that can lead to shape constancy with human observers and computational models.

Can our model be generalized to 3D shapes that are not symmetric? The model can already be applied to nearly symmetric shapes (see Demo 3). For a completely asymmetric shape, the model can be applied as long as there are other 3D constraints that can substantially reduce the family of possible 3D interpretations. Planarity of contours can do the job, as shown with an earlier version of our shape recovery model (Chan et al., 2006). Planarity of contours can often reduce the number of free parameters in 3D shape recovery to just three (Sugihara, 1986). With a three-parameter family of possible interpretations, 3D compactness and minimum surface area are likely to lead to a unique and accurate recovery. This explains why human observers can achieve shape constancy with asymmetric polyhedra, although shape constancy is not as reliable as with symmetric ones.

Finally, we will briefly discuss the psychological and physiological plausibility of one critical aspect of our model. Recall that the model begins with computing a virtual 2D image, which is a mirror reflection of a given 2D real image. Is there evidence suggesting that the human visual system computes 2D virtual images? The answer is "yes." When the observer is presented with a single image of a 3D object, he or she will memorize not only this image, but also its mirror reflection (Vetter, Poggio, and Bülthoff, 1994). Note that a mirror image of a given 2D image of a 3D object is useless unless the 3D object is symmetric. When a 3D object is asymmetric, the virtual 2D image is not a valid image of this object. This means that the observer will never be presented with a virtual image of an object, unless the 3D object is mirror symmetric. When the object is mirror symmetric, then a 2D virtual image is critical, because it makes it possible to recover the 3D shape. The computation of a virtual image has been demonstrated not only in psychophysical but also in electrophysiological experiments (Logothetis, Pauls, and Poggio, 1995; Rollenhagen and Olson, 2000).

# CHAPTER 7

# Noncommutative Field Theory in the Visual Cortex

Alessandro Sarti

Giovanna Citti

## CONTENTS

## INTRODUCTION

We present a new mathematical model of the visual cortex, which takes into account and integrates geometric and probabilistic aspects.

From a geometric point of view, the cortex has been recently described as a noncommutative Lie group, equipped with a sub-Riemannian metric (Hoffman, 1989; Ben Shahar and Zucker, 2003; Bressloff et al., 2001; Citti and Sarti, 2006; Franken et al., 2007; Petitot and Tondut, 1999). The associated Lie algebra is generated by two vector fields that have integral curves in perfect agreement with the shape of association fields of Field, Hayes, and Hess (1993), neurally implemented by the horizontal connectivity of the visual cortex.

The simple cells are able to extract position and orientation of boundaries (i.e., position and momentum variables); therefore, in Sarti, Citti, and Petitot (2008), the structure of the cortex has been identified with a symplectic space, which in turn can be interpreted as the phase space of the retinal plane. The elements of this phase space are operators and, in particular, the generators of the group. Then the cortical manifold is formally equivalent to the phase space in quantum field theory.

A similar point of view was already implicitly taken by Daugmann (1985), when he proposed to use the minima of the classical Heisenberg uncertainty principle to model receptive profiles. Here we prove a similar uncertainty principle, but we directly deduce it from the noncommutativity of the geometrical structure. The minima of the principle are functions able to detect position and orientation with the smallest uncertainty; they are called *coherent states*.

The natural mapping from a world space to its phase space is performed by the Bargmann transform obtained via convolution with the coherent states. Then it will be shown that the action of the simple cells, which is exactly a convolution with the receptive filters, performs such a transform. The norm of the output of the simple cells is generally interpreted as an energy function, always positive, output of the complex cells. Hence, the norm of the Bargmann transform, suitably normalized, will be considered as a probability measure. Consequently, to the image it is associated a natural operator to account for probability distribution, that is, the density operator (Carmichael, 2002).

This approach is in agreement with the probabilistic model proposed by Mumford (1994) and further exploited in August and Zucker (2003); Duits and Franken (2007); Franken, Duits, and ter Haar Romeny (2007);

Williams and Jacobs (1997a); and Zucker (2000). They formulated the assumption that the signal in the cortex can be described as a Markov process. This consideration leads, in turn, to a Fokker–Planck equation in the cortical phase space. Its solution expresses the probability that a point with a specific direction belongs to a contour, and it is implemented by the horizontal connectivity in the three-dimensional (3D) cortical space. The output of the Bargmann transform containing information about image boundaries is propagated by the Fokker–Planck equation, resulting in boundary completion and the filling in of the figure.

Projecting the propagated solution from the 3D cortical space to the retinal plane, we obtain the density operator giving a tensorial representation associated to the perceived image.

In the next section, the neurogeometry of the cortex is presented, following Citti and Sarti (2006). In the third section, we will reinterpret the neurogeometric structure from a probabilistic point of view, by replacing vector fields generators of the Lie algebra by the corresponding operators (i.e., performing the "second quantization").

## THE NEUROGEOMETRICAL STRUCTURE OF THE CORTEX

### The Set of Receptive Profiles and the Lie Group of Rotation and Translation

The retinal plane $M$ will be identified with the two-dimensional (2D) plane $\mathbb{R}^2$ with coordinates $(x,y)$. When a visual stimulus $I(x,y):M \cup \mathbb{R}^2 \rightarrow \mathbb{R}^+$ activates the retinal layer, the cells centered at every point $(x,y)$ of $M$, process in parallel the retinal stimulus with their receptive profile (RP). The RP of a single cell can be modeled as a Gabor filter (see Figure 7.1):

$$\Psi_0(\xi,\eta) = e^{-(\xi^2+\eta^2)}e^{2i\eta} \qquad (7.1)$$

All the other observed profiles can be obtained by translation and rotation of $\Psi_0(\xi,\eta)$ (see Lee, 1996). We will denote $A_{x,y,\theta}$, the action of the group of rotation and translation on $R^2$, which associates to every vector $(\xi,\eta)$ a new vector $(\tilde{\xi},\tilde{\eta})$ according to the following law:

$$(\tilde{\xi},\tilde{\eta}) = A_{x,y,\theta}(\xi,\eta) = \begin{pmatrix} x \\ y \end{pmatrix} + \begin{pmatrix} \cos(\theta) & -\sin(\theta) \\ \sin(\theta) & \cos(\theta) \end{pmatrix} \begin{pmatrix} \xi \\ \eta \end{pmatrix} \qquad (7.2)$$

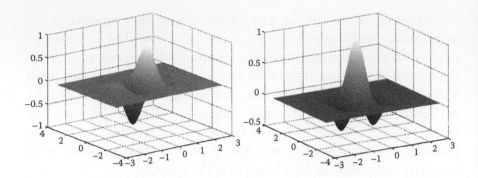

FIGURE 7.1 The real (left) and imaginary (right) parts of a Gabor filter, modeling simple cell receptive profiles.

The action of the group on the set of profiles then becomes

$$\Psi_{x,y,\theta}(\tilde{\xi},\tilde{\eta}) = \Psi_0(A^{-1}_{x,y,\theta}(\tilde{\xi},\tilde{\eta}))$$

### The Action of Simple Cells and the Lie Algebra

The overall output $O$ of the parallel filtering is given by the integral of the signal $I(\xi,\eta)$ times the bank of filters:

$$O(x,y,\theta) = \int_M I(\tilde{\xi},\tilde{\eta})\Psi_{(x,y,\theta)}(\tilde{\xi},\tilde{\eta})d\tilde{\xi}d\tilde{\eta} \qquad (7.3)$$

The selectivity in orientation of the output $O$ is very weak. Several models have been presented to explain the emergence of strong orientation selectivity in the primary visual cortex. Even if the basic mechanism producing strong orientation selectivity is controversial ("push–pull" models [Miller, Kayser, and Priebe, 2001; Priebe et al., 1998], "emergent" models [Nelson, Sur, and Somers, 1995], "recurrent" models [Shelley et al., 2000], to cite only a few), nevertheless, it is evident that the intracortical circuitry is able to filter out all the spurious directions and to strictly keep the direction of maximum response of the simple cells.

For $(x,y)$ fixed, we will denote $\bar{\theta}$ the point of maximal response:

$$\max_\theta \| O(x,y,\theta) \| = \| O(x,y,\bar{\theta}) \| \qquad (7.4)$$

We will then say that the point $(x,y)$ is lifted to the point $(x,y,\bar{\theta})$ (see Figure 7.2). If all the points of the image are lifted in the same way, the level

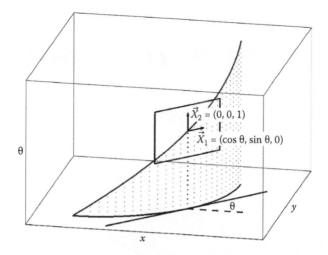

FIGURE 7.2 **(See color insert.)** A two-dimensional curve (in blue) and its three-dimensional cortical lifting in the roto-translation group (in red). The tangent vector to the blue curve is $(\cos(\theta), \sin(\theta))$, so that the tangent vector to its lifted curve lies in the plane generated by $\vec{X}_1$ and $\vec{X}_2$.

lines of the 2D image $I$ are lifted to new curves in the 3D cortical space $(x, y, \bar{\theta})$. The vector $(\cos(\bar{\theta}), \sin(\bar{\theta}))$ is tangent to the level lines of $I$ at the point $(x,y)$, so that the selected value of $\bar{\theta}$ is the orientation of the boundaries of $I$. It has also been proven in Citti and Sarti (2006) that the 3D lifted curves are tangent to the plane generated by the vector fields

$$\vec{X}_1 = (\cos(\theta), \sin(\theta), 0) \quad \vec{X}_2 = (0,0,1) \tag{7.5}$$

### The Noncommutative Structure

We explicitly note that the vector fields $\vec{X}_1\ \vec{X}_1$ are left invariant with respect to the group law of rotations and translations, so that they are the generators of the associated Lie algebra. If we compute the commutator, we obtain the vector $\vec{X}_3$:

$$\vec{X}_3 = [\vec{X}_1, \vec{X}_2] = (-\sin(\theta), \cos(\theta), 0) \tag{7.6}$$

Because it is different from 0, the Lie algebra is not commutative. And, $\vec{X}_3$ is linearly independent of $\vec{X}_1$ and $\vec{X}_2$.

This noncommutative property can be observed starting from the integral curves of the vector fields $\vec{X}_j$. Starting from a point $(0,0,0)$ and moving first in the direction of the vector field $\vec{X}_1$ and then in the direction of

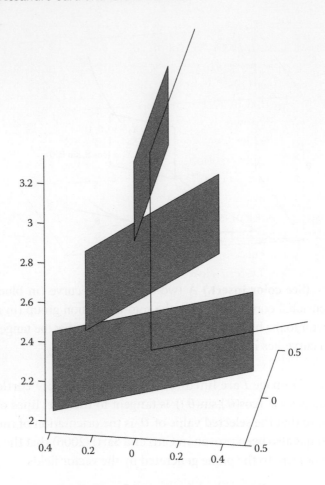

FIGURE 7.3 The composition of two integral curves of the roto-translation group is noncommutative, depending on the order of application of the vector fields.

the vector $\vec{X}_2$, we reach a point $(x, y, \theta)$, different from the point $(x_1, y_1, \theta_1)$ we could reach moving along the vector field $\vec{X}_2$ first and then along the vector $\vec{X}_1$ (see Figure 7.3).

## Association Fields and Integral Curves of the Structure

The natural curves of the structure are the integral curves of the vector fields $\vec{X}_1$ and $\vec{X}_2$, starting at a fixed point $(x_0, y_0, \theta_0)$:

$$\gamma' = (x', y', \theta') = \vec{X}_1(x, y, \theta) + k\vec{X}_2(x, y, \theta) = (\cos(\theta), \sin(\theta), k)$$

$$\gamma(0) = (x_0, y_0, \theta_0) \tag{7.7}$$

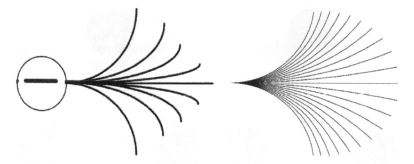

FIGURE 7.4 The association fields of Field, Hayes, and Hess (Field, D.J., Hayes, A., and Hess, R.F., *Vision Res.*, 33, 173–193, 1993) (left) and the integral curves of the vector fields $\vec{X}_1$ and $\vec{X}_2$ with constant coefficients, see Equation 7.7 (right).

and obtained by varying the parameter $k$ in $\mathbb{R}$ (Figure 7.4). These curves can be used to model the local association field as described in Field, Hayes, and Hess (1993) and Gove, Grossberg, and Mingolla (1995).

We can define the distance between two points as the length of the shortest path connecting them. In the Euclidean case, the minimum is obtained within all possible curves, while here we will minimize over the set of integral curves of the vector fields $\vec{X}_1$ and $\vec{X}_2$ (Nagel, Stein, and Wainger, 1985). Using the standard definition, we call length of any curve $\gamma$

$$\lambda(\gamma) = \int \|\gamma'(t)\| \, dt = \int \sqrt{1+k^2} \, dt \tag{7.8}$$

It can be proven that the parameter $k$ expresses the curvature of the projection of the curve $\gamma$ on the plane $(x, y)$ (see Citti and Sarti, 2006). Hence, the geodesics of the group structure, which minimize this quantity, are elastica, as introduced by Mumford (1994) for perceptual completion. Shown in Figure 7.5 is the completion of a Kanitza triangle with curved boundaries by means of the geodesics in the group.

## THE OPERATORIAL STRUCTURE OF THE CORTEX

We will next reinterpret the neurogeometric structure introduced in the first part of the paper from a probabilistic point of view, replacing vector field generators of the Lie algebra by the corresponding operators. This operation is called in physics *second quantization*.

FIGURE 7.5   A Kanitza triangle with curved boundaries (left) and its completion with geodesics in the group (right). The geodesics are not rectilinear, because they minimize the distance in the group (Equation 7.11) that contains the curvature $k$.

## The Cortex as a Phase Space

In Sarti, Citti, and Petitot (2008), we identified the space of simple cells with the phase space of the retinal plane. A simple cell with the receptive field centered at a point $(x, y)$ can be identified with an operator that selects the direction of the boundary at that point. It is then identified with the form

$$\omega = -\sin(\theta)dx + \cos(\theta)dy \qquad (7.9)$$

Note that this is a representation of a general unitary element of the contangent space at the fixed point or, equivalently, of the phase space. Its kernel is a bidimensional plane generated by the first-order operators $X_1$ and $X_2$, with the operatorial version of the vector fields $X_j$, generators of the group and defined in Equation (7.5):

$$X_1 = \cos(\theta)\partial_x + \sin(\theta)\partial_y, \quad X_2 = \partial_\theta$$

These operators describe the propagation along the direction of the vector fields $\vec{X}_1$, $\vec{X}_2$.

These operators, being identified with generators of a vector space, are defined up to a constant. This constant is generally chosen in such a way that the operators are self-adjoint, with respect to the standard scalar product defined on $R^2 \# S^1$, which is defined as follows:

$$<b, h> = \int_0^\pi \int_{R^2} b(x, y, \theta)\overline{h}(x, y, \theta)dxdy\theta$$

where $\overline{h}$ is the complex conjugate of $h$.

Recall that an operator $X$ is self-adjoint if for every couple of functions $b, h$ defined on $\mathrm{R}^2 \times \mathrm{S}^2$,

$$< Xb,h >=< b, Xh >$$

Accordingly, we will choose as generators of the kernel of $\omega$ the operators

$$X_1 = i(\cos(\theta)\partial_x + \sin(\theta)\partial_y), \quad X_2 = i\partial_\theta \qquad (7.10)$$

where $i$ is the imaginary unit.

## The Generators of Rotations and Translations on the Retinal Plane

The two operators $X_1, X_2$, introduced in Equation (7.10) satisfy the noncommutation relation (Equation [7.9]), which defines univocally the Lie algebra of rotation and translation. They act directly on the 3D phase space. However, through the action defined in Equation (7.2), we can obtain their projection on the 2D plane.

We saw that the group acts on the Hilbert space of functions $f$ of two variables in the following way:

$$A_{x,y,\theta}(f(\tilde{\xi},\tilde{\eta})) = f(A^{-1}_{x,y,\theta}(\tilde{\xi},\tilde{\eta}))$$

Since, for every point $(\tilde{\xi},\tilde{\eta})$ fixed, the action $A_{x,y,\theta}$ can be considered a map from the phase space of variables $(x, y, \theta)$ to the 2D retinal plane, and then its differential map sends the operators $X_1, X_2$ to operators $Y_1, Y_2$ on the 2D space:

$$dA(X_1) = Y_1, \quad dA(X_2) = Y_2$$

If $(\xi,\eta) = A^{-1}_{x,y,\theta}(\tilde{\xi},\tilde{\eta})$, simple derivatives show that

$$X_1(f(A^{-1}_{x,y,\theta}(\tilde{\xi},\tilde{\eta}))) = -i(\partial_\xi f)(A^{-1}_{x,y,\theta}(\tilde{\xi},\tilde{\eta})) = -i(\partial_\xi f)(\xi,\eta)$$

and

$$X_2(f(A^{-1}_{x,y,\theta}(\xi,\eta))) = i(\eta\partial_\xi - \xi\partial_\eta)f(\xi,\eta)$$

Hence, we obtain

$$Y_1 = i\partial_\xi \quad Y_2 = i(\eta\partial_\xi - \xi\partial_\eta) \qquad (7.11)$$

The operators $Y_1, Y_2$ can be interpreted as the generators of translation in the direction $\xi$ and rotation around the origin on the 2D retinal plane, respectively. The operators will also be interpreted as position and angular momentum operators, respectively. They satisfy the same commutation rules as the operators $X_1$, $X_2$, as well as the vector fields $\bar{X}_1, \bar{X}_2$ :

$$[Y_1, Y_2] = Y_3, \quad [Y_1, Y_3] = 0 \quad [Y_2, Y_3] = Y_1 \tag{7.12}$$

where

$$Y_3 = \partial_\eta$$

is the generator of translation in the direction orthogonal to $Y_1$, exactly as the corresponding vector $X_3$ is the generator of translations in the direction orthogonal to $X_1$.

The operators $Y_1$ and $Y_2$ represent the propagation along the curves of the bidimensional association fields (see Figure 7.4).

Note that they are self-adjoint with respect to the scalar product in $R^2$:

$$< f, g >_{R^2} = \int_{R^2} f(\xi, \eta) \bar{g}(\xi, \eta) d\xi d\eta$$

## The Uncertainty Principle

Receptive profiles have been interpreted as the minimal of the Heisenberg uncertainty principle by Daugman (1985). His crucial remark is the fact that simple cells are able to detect at the same time position and orientation, but these two quantities do not commute. Hence, the classical uncertainty principle first introduced by Heisenberg in quantum mechanics applies. Roughly speaking, it states that it is not possible to detect exactly both position and momentum and that the variance of their measurements cannot go below a fixed degree of uncertainty. However, there exist functions, the "coherent states," that are able to minimize the degree of uncertainty in both quantities. These minima are Gabor (1946) filters, and Daugman (1985) proved that they are a good model of the receptive profiles of simple cells.

It was proven in Folland (1989) that the uncertainty principle does not apply only to the generators of the Heisenberg group. It applies to any couple of noncommuting operators. Hence, in our setting, it seems more natural to minimize the uncertainty principle for the generators of the Lie algebra of rotations and translations. Because we are looking for functions defined on the 2D space, we will use the representation of the vector fields in terms of the 2D variables, namely, the vector fields $Y_1$ and $Y_2$ introduced in Equation 7.11.

**Proposition 7.1** *The uncertainty principle in terms of these vector fields reads as follows:*

$$|< Y_3 f, f >| \leq 2 \| Y_2 f \| \| Y_1 f \|$$ (7.13)

*where* $\| \; \|$ *is the L2 norm.*

*Proof.* The proof is as follows:

$$|< Y_3 f, f >| = |< (Y_1 Y_2 - Y_2 Y_1) f, f >| =$$ (7.14)

(because the operators are self-adjoint)

$$= |< Y_2 f, Y_1 f > - < Y_1, Y_2 f >| = 2 | Im(< Y_2 f, Y_1 f >)|$$

where *Im* denotes the imaginary part. We conclude the proof, using the Cauchy–Schwartz inequality:

$$2 | Im(< Y_2 f, Y_1 f >)| \leq 2 \| Y_2 f \| \| Y_1 f \|$$ (7.15)

From this inequality and (7.14), the thesis immediately follows.

A slightly more general principle can be obtained if we substitute the norm of $Y_j$ with variance $\| Y_j - a_j \|$, where $a_j$ is the mean value of $Y_j$:

$$|< Y_3 f, f >| \leq 2 \| (Y_1 - a_1) f \| \| (Y_2 - a_2) f \|$$

The products of variances of the position and angular momentum operators have a lower bound, so it is not possible to measure both quantities in an optimal way.

## Coherent States and Receptive Profiles

The coherent states, minima of the uncertainty principle, are the functions that minimize the variance in the measure of position and momentum at the same time.

**Proposition 7.2** *The minimizers of inequality (7.13) satisfy the following equation:*

$$Y_1 u = i \lambda Y_2 u.$$

*Proof.* The proof can be found, for example, in Theorem 1.34 in Folland (1989). It follows from the fact that minimizers must satisfy the equality in the uncertainty principle:

$$|< Y_3 u, u >| = 2\, \|(Y_1 - a_1)u\| \|(Y_2 - a_2)u\|$$

so that also the Schwartz inequality (Equation [7.15]) has to be an equality:

$$| Im(< Y_2 u, Y_1 u >)| = \| Y_2 u \| \| Y_1 u \| \tag{7.16}$$

If the scalar product is equal to the product of the norms, then the vectors $Y_2 u$ and $Y_1 u$ have to be parallel:

$$Y_1 u = i\lambda Y_2 u$$

A direct verification proves that the Gabor filter $\Psi_0$ defined in Equation (7.1) satisfies this equation. As we said before, all the other filters were obtained from this via rotations and translations, so that they are coherent states in the sense of Perelomov (1986). Lee (1996) has shown these states are in agreement with the experimental data of the receptive profile (Figure 7.6).

Let us note that these minimizers are the coherent states of the reducible representation of the group. In Barbieri et al. (2010), we used the irreducible representation of the operators $Y_j$, so that we searched minimizers in the set of functions with fixed frequency in the Fourier domain, and we obtained the classical pinwheels structure, observed experimentaly by Bosking et al. (1997).

## Output of Simple Cells and Bargmann Transform

We recall that the coherent states span the whole Hilbert space of functions defined on $R^2$. They are not a basis of the space, but they form a frame. This

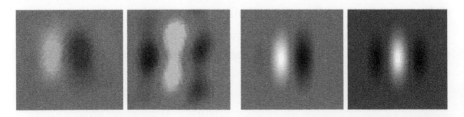

FIGURE 7.6 Receptive profiles of simple cells measured by De Angelis (Reproduced with permission from De Angelis et al., 1995), and the corresponding Gabor functions (right), minimizers of the uncertainty principle (Equation [7.13]).

amounts to saying that any function $f$ on $R^2$ can be represented as a linear combination of coherent states:

$$f(\xi,\eta)=\int C(x,y,\theta)\Psi_{x,y,\theta}(\xi,\eta)dx\,dy\,d\theta$$

but the coefficients $C(x, y, \theta)$ are not unique (Folland, 1989).

The output of single simple cells, defined in Equation (7.3), can be represented as a scalar product:

$$O(x,y,\theta)=<I,\Psi_{x,y,\theta}>$$

and it has the natural meaning of the projection of $I$ in the direction of the specific element $\Psi$ of the frame. Because the functions $\Psi$ are coherent states, the whole output will be identified with a Bargmann transform, which is defined exactly as the scalar product with the entire bank of filters (Antoine, 2000):

$$B_I(x,y,\theta)=\int I(\xi,\eta)\Psi_{x,y,\theta}(\xi,\eta)d\xi\,d\eta \tag{7.17}$$

This transform maps the image $I(\xi, \eta)$ defined on the 2D space, to a function of three variables defined on the phase space. It inherits regularity properties from the properties of the filters. It is simple to verify that

$$X_j B_I(x,y,\theta)=\int I(\xi,\eta)Y_j\Psi_{x,y,\theta}(\xi,\eta)d\xi\,d\eta$$

Due to Proposition 7.2, this implies that the output satisfies

$$X_1 B_I = iX_2 B_I$$

Hence, the Bargmann transform is an entire function in the Cauchy–Riemann (CR) structure generated by $X_1$ and $X_2$ (see Folland, 1989, for the definition of CR structure). In this context, the classical complex derivative

$$\frac{d}{dz}=\partial_x - i\partial_y$$

is replaced by the derivative with respect to the operators $X_j$. Hence, the function $B_I$ output of simple cells is holomorphic in this quasi-complex structure because

$$\frac{d}{dz}B_I = X_1 B_I - iX_2 B_I = 0$$

We quote Kritikos and Cho (1997) to outline the importance of holomorphic functions when studying completion.

### Probability Measure

The norm of the (normalized) Bargmann transform has a probabilistic interpretation. Hence, we can interpret the norm of the output of simple cells as the probability that the image $I$ is in a specific coherent state. More precisely, the probability that the image has a boundary with orientation $\theta$ at the point $(x, y)$ is expressed by

$$P(x, y, \theta) = \| B_I(x, y, \theta) \|$$

Let us explicitly note that the probability is higher if the gradient of $I$ is higher and that this information is neurally provided by the energy output of complex cells.

When in Equation (7.3) we applied the nonmaxima suppression procedure to the function $\| O(x, y, \theta) \|$, for each point $(x, y)$, we assigned a deterministic value $\bar{\theta}$ to the function $\theta$ as the value attaining the highest probability $\| O(x, y, \bar{\theta}) \|$.

### The Density Operator and *P*-Representation

The probability $P$ represents the structure of the image in the phase space. In order to further understand its role, we want to project it back on the retinal 2D space, obtaining the density operator, as defined, for example, in Carmichael (2002):

$$\rho(\xi, \eta, \tilde{\xi}, \tilde{\eta}) = \int \Psi_{x_0, y_0, \theta_0}(\xi, \eta) P(x_0, y_0, \theta_0) \overline{\Psi}_{x_0, y_0, \theta_0}(\tilde{\xi}, \tilde{\eta}) dx_0 \, dy_0 \, d\theta_0 \quad (7.18)$$

The operator is not local, in the sense that it does not depend only on the probability at the point $(x_0, y_0, \theta_0)$, but it is obtained via convolution with the coherent states, defined on noncompact support.

Because

$$\Psi_{x_0, y_0, \theta_0}(\xi, \eta) = \overline{\Psi}_{\xi, \eta, \theta_0}(x_0, y_0)$$

then $\rho$ can be represented as

$$\rho(\xi, \eta, \tilde{\xi}, \tilde{\eta}) = \int \overline{\Psi}_{\xi, \eta, \theta_0}(x_0, y_0) \Psi_{\tilde{\xi}, \tilde{\eta}, \theta_0}(x_0, y_0) P(x_0, y_0, \theta_0) dx_0 \, dy_0 \, d\theta_0$$

which allows an interpretation of $\rho(\xi,\eta,\tilde{\xi},\tilde{\eta})$ in terms of correlations. In fact, the function $\rho$ is the sum over $\theta_0$ of all correlations between the filters centered at points $(\xi,\eta)$ and $(\tilde{\xi},\tilde{\eta})$, weighted by the probability measure $P$.

The function $\rho$ is usually called an *operator*, with an abuse of language, even if it is just the integral kernel of the operator:

$$f \mapsto \int \rho(\xi,\eta,\tilde{\xi},\tilde{\eta}) f(\xi,\eta) d\xi d\eta \qquad (7.19)$$

The representation we provide here is the $P$-representation, because it is a diagonal representation of the operator $\rho$ in *terms* of the coherent states. We explicitly recall that we have to use this diagonal $P$-representation because the coherent states form an overdetermined frame. In this situation, we can apply the following theorem, due to Glauber and Sudarshan:

---

**Theorem 7.1** *The diagonal representation of $\rho$ is invertible in the sense that given* P, *we have the expression of $\rho$, and vice versa, if $\rho$ was known, it could be possible to recover* P *in a unique way.* (See Glauber, 1963, and Sudarshan, 1963, for the proof.)

---

In this sense, the probability $P$ and the density operator $\rho$ are equivalent.

## Tensorial Structure of the Boundaries

In order to understand the meaning of this operator when applied to an image, we will consider the simple case in which receptive profiles centered in $(x_0, y_0)$ and with preferred orientation $\theta_0$ are replaced by 2D vectors $(\cos(\theta_0), \sin(\theta_0))^t$ applied in $(x_0, y_0)$ ($t$ denoting the transposition of vectors). In this way, the previous operator becomes Equation (7.20)

$$\rho(\xi,\eta,\tilde{\xi},\tilde{\eta}) = \int (\cos(\theta_0),\sin(\theta_0))^t \delta_{\xi,\eta}(x_0,y_0) P(x_0,y_0,\theta_0)$$

$$(\cos(\theta_0),\sin(\theta_0))\delta_{\tilde{\xi},\tilde{\eta}}(x_0,y_0) dx_0\, dy_0\, d\theta_0 \qquad (7.20)$$

$$= \int (\cos(\theta_0),\sin(\theta_0))^t P(\xi,\eta,\theta_0)(\cos(\theta_0),\sin(\theta_0))\delta_{\xi,\eta}(\tilde{\xi},\tilde{\eta}) d\theta_0$$

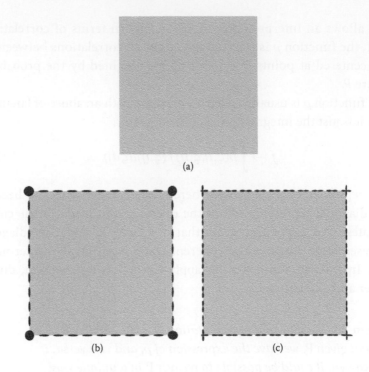

FIGURE 7.7 Boundary representation of a gray square (a) by means of second-order tensors following Equation (7.20) (b) and by means of infinity-order tensors following the density operator Equation (7.18) (c). In this example, the dimensions of the receptive profiles are much smaller than the dimensions of the square.

Hence, $\rho$ is different from 0 only for $(\xi,\eta)=(\tilde{\xi},\tilde{\eta})$, and it has matrix values:

$$\rho(\xi,\eta,\xi,\eta)=\int\begin{pmatrix}\cos^2(\theta_0) & \cos(\theta_0)\sin(\theta_0)\\\cos(\theta_0)\sin(\theta_0) & \sin^2(\theta_0),\end{pmatrix}P(\xi,\eta,\theta_0)d\theta_0$$

With this reduction, we simply associate to every point $(\xi,\eta)$ a rank 2 tensor, which expresses the geometric properties of the boundary and can be considered a second-order approximation of the kernel $\rho$. On the other hand, because $\rho$ defines an operator on the infinity dimensional space of functions, we will interpret the density operator $\rho$ as an infinity-order tensor.

Let us consider, for example, the image of a gray square (see Figure 7.7, top). First, the probability $P$ has been obtained as energy of the Bargmann

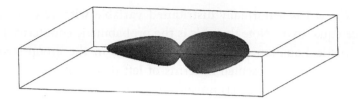

FIGURE 7.8 The fundamental solution of the Fokker–Planck equation in the phase space. An isosurface of intensity is visualized.

transform Equation (7.20), and nonmaximal suppression has been applied. The probability $P$ is 0 far from the boundary. Over each point $(x_0, y_0)$ of the boundary, the probability $P$ is a Dirac Mass in the 3D cortical space, concentrated at the point $(x_0, y_0, \theta_0)$, where $\theta_0$ is the orientation of the boundary at that point, while at the corner $P$, it is the sum of two Dirac Masses.

In this example, the dimension of the coherent states is much smaller than the square. Hence, we can apply the rank 2 tensor approximation. This provides us with a stick along the boundary, and a tensor shaped as a ball at the corners, as in the classical approach of Medioni, Lee, and Tang (2000). In Figure 7.7 (b) it is visualized the second-rank tensor field obtained by Equation (7.20).

In Figure 7.7 (c) it is visualized by the density operator (Equation 7.18). The density operator is one-to-one equivalent to the probability $P$ by the Glauber–Sudarshan theorem, so it keeps all the information contained in the probability density distribution and it corresponds to an infinity rank tensor field. The infinity rank tensorial representation provides us with a stick along the boundary, as in the second-order case, and with an entire cross at the corners, keeping the two distinct directions of the borders.

The density operator is then able to represent arbitrarily complex structures, as we will see in the next sections.

## Propagation of Cortical Activity and the Fokker–Planck Equation

In the section, Association Fields and Integral Curves of the Structure, we observed that different points of the group are connected by the integral curves of the vector fields, cortically implemented by horizontal connectivity. Such connectivity can be modeled in a stochastic setting by the following stochastic differential equation (SDE) first introduced by Mumford (1994) and further discussed by August and Zucker (2003), Williams and Jacobs (1997a), and Sarti and Citti (2010):

$$(x', y', \theta') = (\cos(\theta), \sin(\theta), N(0, \sigma^2)) = \vec{X}_1 + N(0, \sigma^2)\vec{X}_2$$

where $N(0, \sigma^2)$ is a normally distributed variable with zero mean and variance equal to $\sigma^2$. Note that this is the probabilistic counterpart of the deterministic Equation (7.7), naturally defined in the group structure. Both systems are represented in terms of left invariant operators of the Lie group, the first with deterministic curvature and the second with normal random variable curvature. These equations describe the motion of a particle moving with constant speed in a direction randomly changing accordingly with the stochastic process $N$. Let's denote $u$ the probability density to find a particle at the point $(x, y)$ moving with direction $\vec{X}_1$ at the instant of time $t$ conditioned by the fact that it started from a given location with some known velocity. This probability density satisfies a deterministic equation known in literature as the Kolmogorov Forward equation or Fokker–Planck (FP) equation:

$$\partial_t u = X_1 u + \sigma^2 X_{22} u \tag{7.21}$$

In this formulation, the FP equation consists on an advection term in the direction $X_1$, the direction tangent to the path, and a diffusion term on the orientation variable $\theta$ ($X_2$ is the second derivative in direction $\theta$). This equation has been largely used in computer vision and applied to perceptual completion-related problems. It was first used by Williams and Jacobs (1997a) to compute the stochastic completion field; by August and Zucker (2003) and Zucker (2000) to define the curve indicator random field; and more recently, by Duits and Franken (2007) and Franken, Duits, and ter Haar Romeny (2007) applying it to perform contour completion, denoising, and contour enhancement. Its stationary counterpart was proposed in Sarti and Citti (2010) to model the probability of the co-occurence of contours in natural images.

Here we propose to use the FP equation for modeling the weights of horizontal connectivity in primary visual cortex. For this purpose, we are not interested in the propagation in time of $u$, as given by Equation (7.24), but in the fundamental solution of

$$X_1 u(x, y, \theta) + \sigma^2 X_{22} u(x, y, \theta) = \delta(x, y, \theta) \tag{7.22}$$

The fundamental solution of Equation (7.22) is visualized in Figure 7.8, Equation (7.22) is strongly biased in direction $X_1$, and to take into account the symmetry of horizontal connectivity, the model for the probability density propagation has to be symmetrized, for example, considering the sum of the Green functions corresponding to forward and backward FP equations.

This model is in agreement with Sarti and Citti (2010), if we assume that the connectivity is learned by the edge distribution of natural images.

In Figure 7.9, the fundamental solution $\Gamma$ of the Fokker–Planck equation is visualized as second-order tensors (left) and as infinity-order tensor by means of the density operator (right).

## Tensorial Structure of the Image

The probability density $P(x, y, \theta)$, norm of the Bargmann transform and containing information about boundaries of the image, is then propagated by the following equation:

$$X_1 u(x, y, \theta) + \sigma^2 X_{22} u(x, y, \theta) = P(x, y, \theta) \qquad (7.23)$$

where the Dirac delta in Equation (7.22) has been substituted with the forcing term $P(x, y, \theta)$. Equivalently, the distribution $u(x, y, \theta)$ can be obtained by the convolution product:

$$u(x, y, \theta) = \Gamma(x, y, \theta) * P(x, y, \theta)$$

where $\Gamma(x, y, \theta)$ is the Green function, solution of (7.22).

For the square of Figure 7.7, the resulting $u(x, y, \theta)$ fills in the entire figure and structures the inside. In Figure 7.10, a 2D projection of $u(x, y, \theta)$ is visualized by means of a rank 2 tensor field (left) and a rank infinity tensor field corresponding to the density operator (right). In both cases, the tensors induced inside the figure are similar to balls (i.e., they are more isotropic than the tensors in the boundaries). The rank infinity tensor field preserves more information about the global shape of the object. It represents with fidelity the information content of the whole cortical 3D phase space after horizontal connectivity propagation.

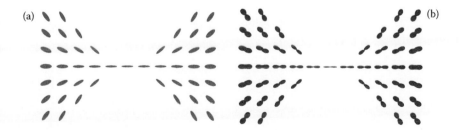

FIGURE 7.9 **(See color insert.)** The fundamental solution $\Gamma$ of the Fokker–Planck equation visualized as second-order tensors (left [a]) and as infinity-order tensors by means of the density operator (right [b]).

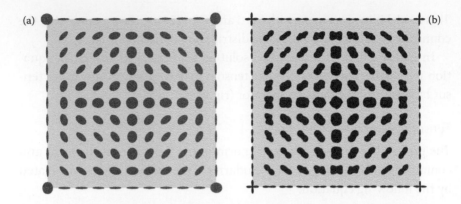

FIGURE 7.10 **(See color insert.)** The inner structure of the square obtained after propagating the lifted boundaries via the Fokker–Planck fundamental solution. The probability density is visualized as second-order tensors (left [a]) and as infinity-order tensor by means of the density operator (right [b]).

## ACKNOWLEDGMENT

The authors are grateful to G. Sanguinetti, who helped to generate some of the figures presented in the text.

# CHAPTER 8

# Contour-, Surface-, and Object-Related Coding in the Visual Cortex

Rüdiger von der Heydt

## CONTENTS

## INTRODUCTION

At the early processing stages in visual cortex, information is laid out in the form of maps of the retinal image. However, contrary to intuition, uniform surfaces are not mapped by uniform distributions of neural activity. We can perceive the three-dimensional (3D) shape of a uniform surface, but stereoscopic neurons are activated only by the contours of the surface, not the uniform interior. We perceive a uniform color figure, but color-selective neurons respond about five times less to the interior of the figure than to its boundaries. How cortical neurons signal contour and surface features is well known, but we do not yet understand how the brain "organizes" these feature signals to represent surfaces and objects. In this

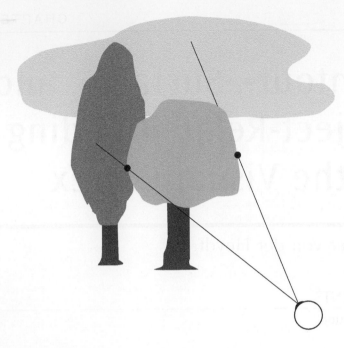

FIGURE 8.1 The problem of image segmentation and the definition of occluding contours.

chapter, I summarize studies showing that the visual cortex codes surface color, depth ordering of surfaces, and border ownership in the contour signals. The signaling of border ownership (the one-sided assignment of borders that determines perceptual figure–ground organization) is the key finding, because it reveals mechanisms of feature grouping. I discuss how these mechanisms might be used by the system to represent objects and compute surface attributes.

Biological vision systems infer 3D structure from images and have evolved to perform this task in a world that has a specific physical structure. The fundamental problem of vision is that 3D scenes are projected onto a two-dimensional (2D) receptor surface; things that are widely separated in space give rise to adjacent regions in the projection (Figure 8.1). The resulting images are composed of regions corresponding to different objects. The boundaries between the regions are given by the geometry of interposition and are called *occluding contours*. Thus, theoretically, images of 3D scenes can be decomposed, or "segmented" into regions separated by contours. The segmentation result might then be used to infer physical surfaces and the shapes of objects. To do this in practice is difficult, and the primate visual system devotes hundreds of millions of cells to this task.

## RESULTS

### Surface Color[1]

One of the most surprising findings of modern vision research was the discovery of orientation selectivity of neurons in the visual cortex and the emphasis on edges (discontinuities of intensity and color) in the cortical representation. This can be appreciated by looking at the profile of neuronal activity across the representation of the surface of a uniformly colored object (Figure 8.2). The results presented here are based on single-cell recordings from the macaque visual cortex under awake behavioral conditions (Friedman, Zhou, and von der Heydt, 2003). The curves show the firing rates of the neurons as a function of the location of the receptive field (RF) relative to a square figure. The curves represent averaged responses of neurons with near-foveal receptive fields; the square measured either 4° or 6° on a side, a multiple of the size of the receptive fields. About 80% of the neurons in V2 and the supragranular layers of V1 are strongly orientation selective. As shown by the top curves in Figure 8.1, these cells respond to the edges but are virtually unresponsive to the uniform interior of the square. The curves below show that nonoriented and weakly oriented cells, although activated by the uniform surface, also emphasize the edges by about a factor of 3. As a result, the total activity in these regions of cortex is about five times higher at the edges than in the center of the square (bottom curve; in combining the activity profiles of oriented and nonoriented cells, the relative number of cells was taken into account along with the fact that only a small fraction of the oriented cells are activated by the edges at any given orientation, while all nonoriented cells contribute activity).

An important point is that the proportion of color-selective neurons is at least as high among oriented cells as it is among nonoriented cells (Friedman, Zhou, and von der Heydt, 2003). Because of the overwhelming preponderance of orientation-selective neurons, this means that color information is predominantly carried by orientation-selective edge responses.

How can edge signals code for surface color? Edge-selective neurons respond when their receptive field is on the border between two regions of different color. If the neurons were to signal the color of a surface, and not just the presence of an edge, they should be able to differentiate on which side the color is located. We found that most cells discriminate the

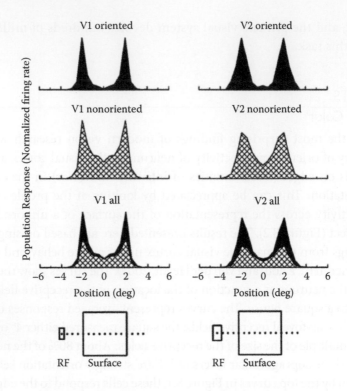

FIGURE 8.2 Response profiles across the representation of a square of uniform color in the macaque visual cortex. The plots show the averaged population responses for orientation-selective cells (top), nonorientation-selective cells (below), and for the combined activity of both (bottom). (Left) area V1 (supragranular cells only); (right) area V2. The responses are plotted as a function of the receptive field position relative to the square, as depicted schematically at the bottom. The color and the orientation of the square were optimized for each cell. The combined responses (bottom plot) were calculated by weighting the groups according to their respective encounter frequencies and taking into account that the expected firing rates of oriented cells are only about 20% of their maximum firing rates, because a figure of a given orientation will stimulate most cells at nonoptimal orientation. (Reprinted with permission from Friedman et al., 2003.)

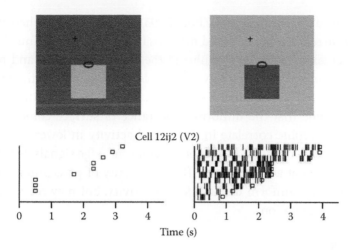

FIGURE 8.3 Selectivity for edge contrast polarity. Responses of an example neuron of V2. The raster plots show the sequences of action potentials in response to repeated presentations of the stimuli depicted above (oval indicates receptive field). Small squares mark the end of the monkey's fixation period. (Modified from Zhou et al., 2000. With permission)

polarity of the color difference at the edge. Figure 8.3 shows an example of such a neuron. (The response profiles of Figure 8.2 do not reveal selectivity for contrast polarity, because responses were averaged over neurons without regard to contrast preference.) Thus, the edge responses of oriented neurons carry directional color information that is essential for coding surface color.

The strong dominance of edge signals in the cortex is hard to reconcile with our subjective experience when looking at a surface of uniform color; the color is no less vivid in the center than at the edges. An attempt to resolve this paradox is the "filling-in" theory that postulates a representation in which color signals spread from the boundary of a surface into the interior and fill it up, thus creating the uniform distribution that perception suggests. We wondered if perhaps a subpopulation of V1 or V2 neurons participates in this process, and that some of the signals that we recorded for the center of the square (Figure 8.2) might reflect filling-in. Thus, we studied a paradigm of illusory filling-in. We measured perceptual filling-in in monkeys and recorded the neural signals in the visual cortex of the same animals (Friedman, Zhou, and von der Heydt, 1999;

von der Heydt, Friedman, and Zhou, 2003). We found that when the fill-ing-in occurred, the nonoriented neurons in V1 and V2 continued to sig-nal the actual color of the stimulus at their receptive fields, and not the color that was perceived. However, the color edge signals of the oriented neurons decayed at a rate that was consistent with the filling-in.

We conclude that the uniform appearance of surface color does not have an isomorphic correlate in the neural activity in lower-level visual areas. Surface color might be computed from the edge signals, which pre-sumably occurs at a higher level. The uniformity of perception does not correspond to a uniform distribution of activity, but may reflect a more abstract representation.

## Surface Depth

Color information is directly available at the receptor level, but depth is the dimension that is lost in the image and needs to be reconstructed by the system. Primates and many other species have a specific mecha-nism for this, using the disparity between the images of the two eyes, providing binocular stereopsis. Because disparity information derives from matching structures in two different images, which cannot be done in uniform image regions, stereoscopic information is charac-teristically sparse. In the case of a uniform square, for example, ste-reoscopic information is available from the edges but not the surface. Disparity can define the depth of the contour of the square but does not indicate whether the surface inside the contour is flat, convex, or concave, and if it is located at the depth of the contour or behind the contour. It could be an object in front, or a surface in back that is seen through a square window.

The sparse representation of depth in the cortex is illustrated in Figure 8.4, where the top row shows the response profiles of three disparity selective neurons. In each case, the stimulus, a uniform square, was pre-sented with the optimal disparity for the neuron. As a result, the edges of the square produced strong responses. However, the surface did not elicit any activity, as expected, because there was no structure in the receptive field that could stimulate the disparity selective-response mechanism.

A different situation is illustrated in the second row of Figure 8.4. It shows the responses of the same neurons to random-dot stereograms, portraying the square in depth, floating in front of a background sur-face. (This kind of stimulus in which a figure is defined by disparity but

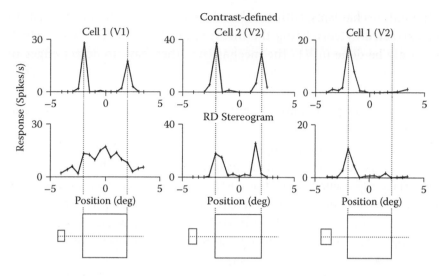

FIGURE 8.4 The response profiles across luminance-defined and disparity-defined figures of three disparity-selective cells in macaque visual cortex. Conventions are as in Figure 8.3. (Top) The luminance-defined figures produced edge-selective responses in all three cells, as expected, because the cells were orientation selective. (Below) The random-dot stereograms produced surface response in cell 1, but edge-selective responses in cells 2 and 3. Stereoscopic edge selectivity was found only in V2. (Reprinted with permission from Von der Heydt et al., 2000.)

has no contours in monocular view, has been called "cyclopean" [Julesz, 1971].) As in the case of the uniform figures, the squares were presented with the optimal disparity for each neuron. Thus, it might be expected that the neurons would now be activated whenever their receptive field was inside the square. This was the case for neurons of V1, as can be seen in the example of cell 1. However, neurons of V2 often responded again in the edge-selective manner, producing no activity when the receptive field was inside the cyclopean square, despite the fact that the random-dot texture had the optimal disparity (von der Heydt, Zhou, and Friedman, 2000).

It is amazing that neurons in V2 can detect the contours of cyclopean figures. These neurons respond to color-defined edges, as do simple and complex cells of V1, and also respond to cyclopean edges that are devoid of edge contrast. Clearly, these two conditions require radically different mechanisms. Unlike color, disparity information first needs to be extracted by correlating the information from the two eyes, which can only be done in V1. The mechanisms then have to detect edges in the disparity map. The cyclopean edge cells of V2 are also exquisitely tuned to the orientation of the 3D edge, and the tuning generally agrees with their tuning for contrast-defined edges (von der Heydt, Zhou, and Friedman, 2000).

The presence of stereoscopic surface responses in V1 and the emergence of edge selectivity in V2 reveal something general about the strategy of surface representation in the cortex. Random-dot stereograms carry ample disparity information all across the area of the displayed object and thus define the 3D shape of its surface perfectly. And yet, the system shifts the emphasis from surface to edge signals. The emergence of a 3D edge representation with orientation-selective neurons in V2 is analogous to the emergence of orientation selectivity in simple cells of V1. The repetition of the same strategy in the processing of depth indicates that representation of contours plays a fundamental role, not only in 2D, but also in the representation of the 3D shape.

## Border Ownership

We can see from Figure 8.1 that the identification of occluding contours is important for two reasons. First, they separate features of different objects, which have to be kept separate in the processing. Second, they carry information about the shape of the occluding object. Although an occluding contour is also the boundary to a background region, it should not be processed with that region; the shape of the background region is an accidental product of the situation of occlusion (Nakayama, Shimojo, and Silverman, 1989). Thus, the identification of occluding contours involves two tasks: finding the borders and assigning them to the correct side, the foreground side.

For detection, a variety of cues are available. A color discontinuity is probably the most reliable indicator; because foreground and background are usually unrelated, there is a high chance that they differ substantially in color. This is exploited by simple and complex cells that effectively extract contrast borders (see Figure 8.2 top). Another possibility is detecting discontinuity of depth, which is illustrated in Figure 8.4 (cells 2 and 3). Relative motion and dynamic occlusion are important cues, too.

The second task, assigning the contours to the foreground, requires a somewhat different repertoire of mechanisms. Discontinuity of color, which is powerful for detection of contours, provides no clues as to which side is

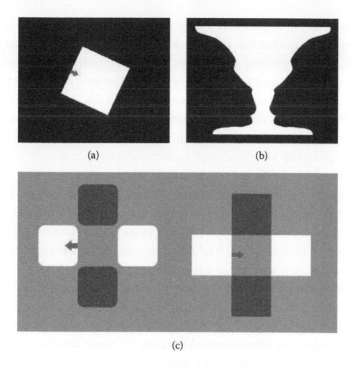

(a)　　　　　　　　　(b)

(c)

FIGURE 8.5 Assignment of "border ownership" in perception. (a) Physically, the square is just a region of higher luminance surrounded by a region of lower luminance, but perceptually, it is an object, and the border between the regions is "owned" by the object. (b) A figure in which border ownership is ambiguous. (c) On the left, four squares are perceived as four different objects, whereas similar squares on the right are perceived as two crossbars, one of which appears transparent. Note the different assignment of border ownership (arrows).

foreground and which is background. The border between a gray and a brown region could be either the contour of a gray object on one side or the contour of a brown object on the other side. Discontinuity of depth, on the other hand, is a perfect cue because it tells which side is in front. Cells that respond to edges in random-dot stereograms are usually (75%) selective for the polarity of the edge (depth ordering of surfaces). An example is cell 3 in Figure 8.4, which responded to the far-near edge in the random-dot stereogram (left-hand edge of square) but not to the near-far edge (right-hand edge of square).

Gestalt psychologists made the surprising observation that the visual system assigns borders also in the absence of unequivocal cues such as disparity in a random-dot stereogram. In Figure 8.5a, the white region is perceived as an object and the black-white boundary as its contour. Apparently, the system has rules to decide which regions are likely to be objects and which ground, and applies these "automatically." This compulsion to assign figure and ground is nicely demonstrated by Rubin's vase figure (Rubin, 1921). My version is shown in Figure 8.5b. Here, different rules seem to be in conflict, and the assignment flips back and forth between the two alternatives, which both lead to perceptions of familiar shapes.

A change in the assignment of border ownership not only affects the depth ordering of surfaces, it can also imply a restructuring of the perceived objects. Figure 8.5c shows, on the left, four squares with rounded corners. On the right, the same squares are shown without the rounding. It can be seen that perception reorganizes the regions: two bars are now perceived instead of the four squares, and the center region that was ground before, now appears as a transparent overlay.

The perceptual interpretations demonstrated in Figure 8.5 obviously involve the image context. For example, the edges marked by arrows in Figure 8.5c are locally identical in both configurations, but the assignment differs according to the context. Alteration of some details elsewhere in the display led to a new interpretation.

The concepts of border ownership and figure–ground have been used in parallel in the literature to describe those perceptual phenomena. The power of the border ownership concept in interpreting and modeling the perceptual observations was recognized (Nakayama, Shimojo, and Silverman, 1989; Sajda and Finkel, 1995), but how the brain codes border ownership, or figure–ground, has long been a mystery. Sajda and Finkel speculated that synchronous oscillations might be the vehicle of the neural coding. However, recent single-cell recordings have suggested a different answer that is surprisingly simple (Zhou, Friedman, and von der Heydt, 2000). The key finding is that individual neurons have a (fixed) border ownership preference, and that each piece of contour is represented by two groups of neurons with opposite border ownership preferences (Figure 8.6). Thus, the differential activity of the two groups signals the direction of border ownership.

Support for this view is provided, among other observations, by a comparison of the neuronal responses to color-defined figures and random-dot stereograms (Figure 8.7). The raster plots show the responses of a V2 neuron under eight different conditions as illustrated by the stimulus cartoons that should be interpreted as perspective drawings of the stimulus configurations. The four conditions at the top (a,b,c,d) are the tests with color-defined figures: figure locations left and right of the receptive field are illustrated

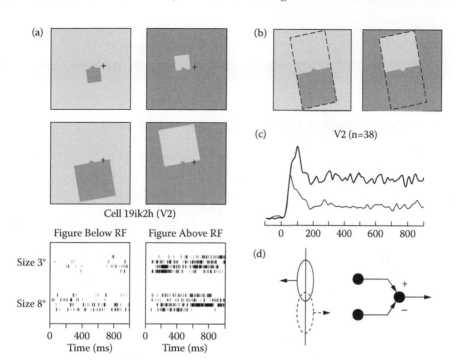

FIGURE 8.6 Border ownership selectivity in neurons of the visual cortex. (a) Example neuron of area V2. Conventions are as in Figure 8.3. Note the response difference between conditions illustrated left and right. The left and right stimuli are locally identical, as shown in (b). (c) Time course of averaged responses of V2 neurons; thick line, preferred side of figure; thin line, nonpreferred side. Each neuron was also tested with displays of reversed contrast (not illustrated), and responses for both contrast polarities were averaged. (d) The differential response between neurons with opposite side preference is thought to signal border ownership. (Modified from Zhou et al., 2000.)

in left and right columns; two rows, A-B and C-D, show tests with the two contrast polarities. The bottom part shows the stereoscopic test. Again, left and right columns show left and right figure locations, but in this test, the "figure" was either a square floating in front of a background (e,h), or a window through which a surface in the back was visible (f,g). It can be seen that the neuron preferred figure-left for the color-defined displays. For stereograms, it responded in the conditions where the front surface was to the left of the receptive field (e,f) but was silent when the front surface was to the right (g,h). Thus, with stereograms, which define depth order of surfaces

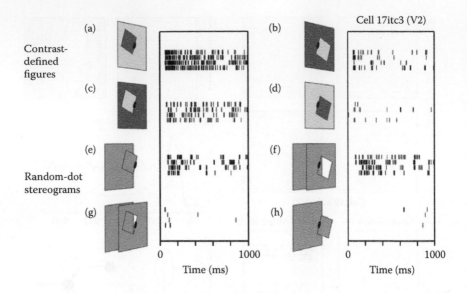

FIGURE 8.7 Responses of an example neuron to contours of luminance-defined figures (a,b,c,d) and to disparity-defined figures (e,f,g,h). The depth configurations are illustrated schematically, by depicting the square figures (e,h) and windows (f,g) in perspective. Oval indicates projection of receptive field. (Reprinted with permission from Qiu and von der Heydt, 2005.)

unequivocally, the neuron responded according to the stereoscopic edge in its receptive field and was selective for left border ownership. Under these conditions, the location of the square shape was irrelevant: the edge could be the right-hand edge of a square figure (e), or the left-hand edge of a square window (f). But with color-defined displays, the neuron responded according to the location of the shape: figure-left produced stronger responses than figure-right. Thus, in the absence of the disparity cue, the neuron "assumes" that the square is an object and the surrounding region is the background. This is exactly how we perceive the white square in Figure 8.5a.

The reorganization of surfaces demonstrated in Figure 8.5c can also be observed in the neuronal responses of V2, as illustrated in Figure 8.8 (Qiu and von der Heydt, 2007). We compared displays of a single square (Figure 8.8a) with the transparent display (Figure 8.8b) and a display of four squares with rounded corners which do not produce perception of transparency (Figure 8.8c). Note that in the top row of displays, the edge under the receptive field (ellipse) is owned left in (a) and (c), but right in (b). The ownership is opposite in the displays below. The curves show the time course of

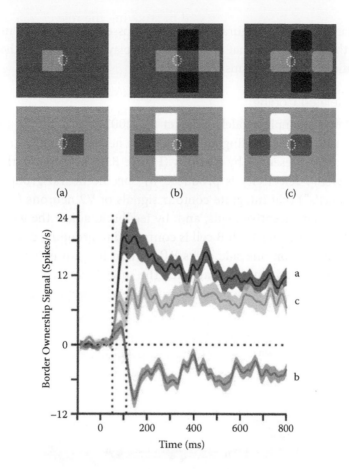

FIGURE 8.8 Border ownership signals parallel perceptual reorganization in transparency displays. In (a) and (c), top, the border marked by a dashed oval is owned left, but in display (b), top, it is owned right. In the displays below, ownership is reversed. The local edge is identical in all six displays. The curves show the corresponding border ownership signals (difference between responses to displays at top and below); average of 127 cells, error bands indicate standard error of the mean (SEM). (Modified from Qiu and von der Heydt, 2007. With permission.)

the averaged border ownership signals of V2 neurons. It can be seen that the signal for (b) was reversed compared to (a) and (c). This shows that the neurons assign border ownership(BOS) according to the transparent interpretation. Particularly striking is the difference between the responses to (b) and (c), which differ only in the presence/absence of "X-junctions." The signal reversal in *b* occurred with a slight delay (dotted vertical lines). These results show that V2 codes configuration *b* as two crossed bars rather than representing the four squares that make up the display. It reorganizes the visual information in terms of plausible objects.

## Object-Related Coding

The following simple model (Craft et al., 2007) explains our findings on border ownership coding and might also help us to understand how surfaces are represented by neurons (Figure 8.9). It proposes that border ownership selectivity is produced by specific neural grouping circuits ("G cells") that integrate contour signals of V2 neurons ("B cells") with cocircular receptive fields, and, by feedback, adjust the gain of the B cells. In this scheme, each B cell is connected to grouping circuits integrating contours on one side of its RF, so that its gain is increased if a convex shape is present on that side. The model assumes G cells with integration fields of various sizes according to a scale pyramid. It predicts

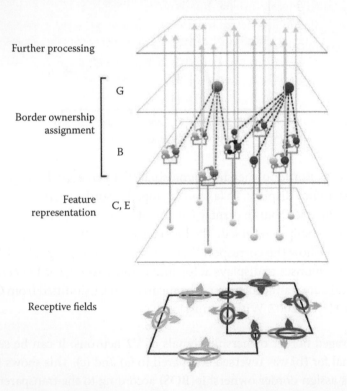

FIGURE 8.9 **(See color insert.)** Neural network model of border ownership coding. See text for further explanation. Reproduced, with permission, from Craft, Schutze, Niebur, and von der Heydt (2007).

border ownership assignment correctly for the various conditions. B cells model border ownership selective neurons as recorded in V2. G cells are assumed to reside in a higher-level area such as V4. As yet, their existence is hypothetical.

One attraction of this model is that it describes not only figure–ground organization but can also be extended to explain how the resulting figure representations are accessed and selected by top-down attention. We assume that top-down attention excites the G cells at the focus of attention (Figure 8.10a,b). Because excitation of a G cell turns up the gain of the B cells that are connected with that G cell, the contour signals representing the figure are enhanced. The illustration shows how the occluding figure (Figure 8.10a) or the underlying figure (Figure 8.10b) can be selected. The latter cannot be accomplished with a spatial attention model, because it

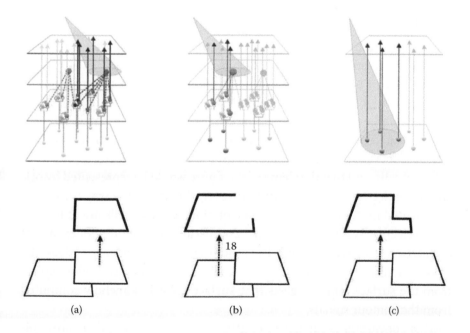

(a)     (b)     (c)

FIGURE 8.10 **(See color insert.)** Explaining selective attention by the model of Figure 8.9. It is assumed that volitional ("top-down") attention excites neurons in the G cell layer as illustrated by a yellow spotlight. In this model, attention enhances the correct contour segments, whether foreground (a) or background objects (b) are attended. In contrast, a spatial attention model extracts a mixture of contours from both foreground and background objects (c). (Modified from Craft et al., 2007. With permission.)

pools the occluding contour with the contours of the underlying figure (Figure 8.10c).

The model predicts that top-down attention should have an asymmetrical effect on the B cells: Because each B cell is connected only to G cells on one side (its preferred border ownership side), the responses of a B cell are enhanced only if the focus of attention is on that side. Experiments in which the monkey attended to a figure on one side or the other confirmed this prediction. The attention effect was asymmetrical, and the side of attentive enhancement was generally the same as the preferred side of border ownership (Qiu, Sugihara, and von der Heydt, 2007). Thus, the model accounts for three findings. It explains how the system uses image context to generate border-ownership signals, it explains the spatial asymmetry of the attention influence, and it explains why the side of attention enhancement is generally the same as the preferred side of border ownership.

## DISCUSSION

We started out by considering the population response profile across the cortical representation of a uniform figure on a uniform background, noting that in both V1 and V2 the contours of the figure are strongly enhanced relative to the interior of the figure. Orientation-selective cells respond exclusively to the contours, whereas nonoriented cells are also activated by the uniform interior. One might expect that a surface attribute like color would be represented mainly by color-selective non-oriented cells, and that the shape of the figure would be represented by orientation-selective noncolor cells. However, this is not the way things are represented in the visual cortex. Orientation-selective cells are at least as color selective as nonoriented cells, and because of the preponderance of orientation-selective cells and the response enhancement at the edges, the color signals from the contours are much stronger than the color signals from the surface. This suggests that surface color is somehow computed from the contour signals.

This conclusion is further strengthened by the finding that most color-and-orientation cells signal the direction of the color gradient at the contour. Again, this is contrary to what is commonly assumed—namely, that selectivity for contrast polarity is found only in simple cells. This is clearly not the case in the primate visual cortex. Simple cells are much less frequent in V2 than in V1 (Levitt, Kiper, and Movshon, 1994), but selectivity for contrast polarity is undiminished (Friedman, Zhou, and von der Heydt, 2003).

We saw a similar transition of coding in the case of depth: In V2, 3D edge selectivity emerges, and cells respond to the contours of cyclopean figures. Here also cells are generally selective for edge polarity, which means that they signal the depth ordering of surfaces at the contour. This means that these neurons signal edges not only for the purpose of coding shape, for which position and orientation of edges would be sufficient, but also for representing the surface. Perhaps the 3D shape of the surface is also derived from the contour representation.

Border ownership coding is the string that ties a number of diverse findings together. It can help us to understand how the visual cortex imposes a structure on the incoming sensory information and how central mechanisms use this structure for selective processing. Border ownership is about relating contours to regions. The most interesting aspect is the global shape processing that is evident from the differential responses to displays that are identical within a large region around the receptive field (Figure 8.6a,b). It implies that the four sides of the square are assigned to a common region, which is a form of "binding," the essence of creating object representations. In our model, we propose a mechanism of two reciprocal types of connectivity, a converging scheme for the integration of feature signals, and a diverging scheme of feedback connections that set the gain of the feature neurons. This gain change is what leads to the observed different firing rates for figures on preferred and nonpreferred sides. Besides this effect that is caused by the visual stimulus, the feedback connections serve in top-down attention for the selective enhancement of the feature signals (B cells) representing the attended object. A large proportion of V2 neurons shows this dual influence of stimulus-driven grouping and top-down attentive modulation (Qiu, Sugihara, and von der Heydt, 2007). Moreover, the neurons show an asymmetry of attentive modulation that is correlated with the side of border ownership preference, as predicted by the model.

Our model differs from many other models of perceptual organization in that it postulates a dedicated neural circuitry for the grouping. The grouping in our model is not the result of self-organization of activity within the cortical area (which would be limited by the length and conduction velocity of horizontal connections in the cortex), but is produced by an extra set of neurons, the G cells. We assume that these cells are not part of the pattern recognition pathway, but specifically serve to provide a structure onto which central processes can latch, forming an interface between feature representation and central cognitive mechanisms. This specificity of function is an important difference: The G cells with their relatively large receptive fields do not have the high resolution required for form recognition. Therefore, the system needs only a relatively small number of G cells to cover the visual space.

I suggest that this general scheme could also explain how the brain represents surfaces. A surface representation might correspond to a set of G cells that the system activates when a task requires specific surface information. Take the example of hiking in the mountains. Here, the visual system needs to provide information about the position of the object to be stepped on, such as a rock, and the surface normal at the point to be touched by the foot. The calculation of the surface normal would involve the activation of a cluster of G cells corresponding to the location of the rock, and these cells would then selectively enhance the signals representing the contours of the rock. The enhanced contour signals would then be collected by a "surface normal calculator" downstream in the pathway. Neurons that represent the 3D orientation of lines exist in V4 (Hinkle and Connor, 2002), and from the 3D orientations of the contour elements, the system could calculate the approximate orientation and curvature of the surface and its normal. A similar process can be conceived for the computation of surface color from contour signals. Thus, I envision surface representation more like a procedure the system can call when needed than a pattern of neural activity representing the surface points in space.

## ENDNOTE

1 "Color" in this paper refers to both chromatic and luminance dimensions. Thus, I include gray values among the colors.

# CHAPTER 9

# Visual Surface Encoding

## A Neuroanalytic Approach

Christopher W. Tyler
Lora T. Likova

## CONTENTS

## DEMAND CHARACTERISTICS OF VISUAL ENCODING

A primary goal of visual encoding is to determine the nature and motion of the objects in the surrounding environment. In order to plan and coordinate actions, we need a functional representation of the scene layout and of the spatial configuration and the dynamics of the objects within it both in the picture plane and in depth. The properties of the visual array, however, have a metric structure entirely different from those of the spatial configuration of the objects. Physically, objects consist of aggregates of particles that cohere together. Objects may be rigid or flexible, but in either case, an invariant set of particles is connected to form a given object. The visual cues (such as edges or gradients) conveying the presence of objects to the brain or to artificial sensing systems, however, share none of these properties. The visual cues conveying the presence of objects may change in luminance or color, and they may be disrupted by reflections or occlusion by intervening objects as the view point changes. The various cues such as shading, texture, color binocular disparity, edge structure, and motion vector fields may even carry *inconsistent* object information. As exemplified by the dodecahedron drawn by Leonardo da Vinci (Figure 9.1a), any of these cues may be *sparse*, with a lack of information about the object structure across gaps where cues are missing. Also, any cue may be *variable* over time. Nevertheless, despite the sparse, inconsistent, and variable nature of the local cues, we perceive solid, three-dimensional (3D) objects by interpolating the sparse depth cues into coherent spatial structures generally matching the physical configuration of the objects.

In the more restricted domain of the *surface structure* of objects in the world, surfaces are perceived not just as flat planes in two dimensions, but as complex manifolds in three dimensions. Here we are using "manifold" in the sense of a continuous two-dimensional (2D) subspace of the 3D Euclidean space. A compelling example of perceptual surface completion,

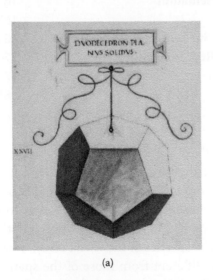

(a)                    (b)

FIGURE 9.1 **(See color insert.)** (a) Dodecahedron drawn by Leonardo da Vinci (1509). (b) Volumetric completion of the white surface of a cylinder (Reprinted with permission from Tse, 1999.) Note the strong three-dimensional percepts generated by the sparse depth cues in both images.

developed by Tse (1999), depicts an amorphous shape wrapping a white space that gives the immediate impression of a 3D cylinder filling the space between the curved shapes (Figure 9.1b). Within the enclosed white area in this figure, there is no information, either monocular (shading, texture gradient, etc.) or binocular (disparity gradient), about the object structure. Yet our perceptual system performs a compelling reconstruction of the 3D shape of a cylinder, based on the monocular cues of the black border shapes. This example illustrates the flexibility of the surface-completion mechanism in adapting to the variety of unexpected demands for shape reconstruction.

It is important to stress that such monocular depth cues as those in Figure 9.1 provide a perceptually valid sense of depth and encourage the view that the 3D surface representation is the *primary* cue to object structure (Likova and Tyler, 2003; Tyler, 2006). Objects in the world are typically defined by contours and local features separated by featureless regions (such as the design printed on a beach ball, or the smooth skin between facial features). Surface representation is an important stage in the visual coding from image registration to object identification. It is very unusual to experience objects (even transparent objects), either tactilely or visually, except through their surfaces. Developing a means of representing the proliferation of surfaces before us is therefore a key stage in the neural processing of objects.

## THEORETICAL ANALYSIS OF SURFACE REPRESENTATION

The previous sections indicate that surface reconstruction is a key factor in the process of making perceptual sense of visual images of 2D shapes, which requires a midlevel encoding process that can be fully implemented in neural hardware. A full understanding of neural implementability, however, requires the development of a quantitative simulation of the process using neurally plausible computational elements. Only when such a computation has been achieved can we be assured that the subtleties of the required processing have been understood.

Surface reconstruction has been implemented by Sarti, Malladi, and Sethian (2000) and by Grossberg, Kuhlmann, and Mingolla (2007) as a midlevel process operating to coordinate spatial and object-based attention into view-invariant representations of objects. The basic idea is to condition the three types of inputs as sparse local depth signals into the neural network, and allow the output to take the form of a *surface* in egocentric

3D space rather than as object reflectance or attentional salience in the 2D space of the visual field. Figure 9.2 illustrates that, to match the characteristics of human visual processing, such an implementation must conform to the stereoscopic limits on the curvature of continuous surfaces discovered by Tyler (1973, 1991). This surface smoothness limit has now been validated by Nienborg et al. (2004) in cortical cells, indicating the relevance of a surface-level reconstruction of the depth structure even in primary visual cortex. Once the depth structure is defined for each visual modality alone (e.g., luminance, disparity, motion), the integration of the corresponding cues should be modeled by Bayesian methods in relation to the estimated reliability of each cue to determine a singular estimate of the overall 3D structure in the visual field confronting the observer.

The concept of surface representation requires a surface *interpolation* mechanism to represent the surface in regions of the field where the information is undefined. Such interpolation is analogous to the "shrink-wrapping" of a membrane around an irregular object such as a piece of furniture.

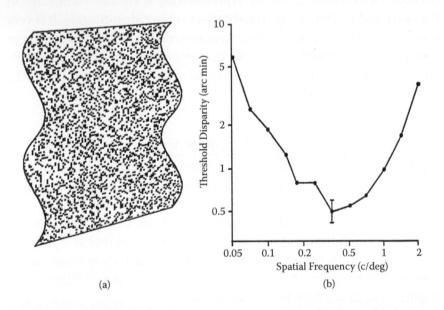

(a)                                      (b)

FIGURE 9.2   (a) Depiction of a random-dot surface with stereoscopic ripples. (b) Thresholds for detecting stereoscopic depth ripples, as a function of the spatial frequency of the ripples. (From Tyler, 1991. With permission.). Peak sensitivity (lowest thresholds) occurs at the low value of 0.4 cycle/deg (2.5 deg/cycle), implying that stereoscopic processing involves considerable smoothing relative to contrast processing.

Whereas a (2D) receptive-field summation mechanism shows a stronger response as the amount of stimulus information increases, the characteristic of an *interpolation* mechanism is to *increase* its response, or at least not decrease its response, as stimulus information is *reduced* and more extended interpolation is required. (Of course, a point will be reached where the interpolation fails and the response ultimately drops to zero.)

Once the object surfaces have been identified, we are brought to the issue of the localization of the object features relative to each other, and relative to those in other objects. Localization is particularly complicated under conditions where the objects could be considered as "sampled" by overlapping noise or partial occlusion—the tiger behind the trees, the face behind the window curtain. However, the visual system allows remarkably precise localization, even when the stimuli have poorly defined features and edges (Toet and Koenderink, 1988; Kontsevich and Tyler, 1998; Likova and Tyler, 2003). Furthermore, sample spacing is a critical parameter for an adequate theory of localization. Specifically, no low-level filter integration can account for interpolation behavior beyond the tiny range of 2 to 3 arc min in foveal vision (Morgan and Watt, 1982), although the edge features of typical objects, such as the form of a face or the edges of a computer monitor, may be separated by blank regions of many degrees. Thus, the interpolation required for specifying the shape of most objects is well beyond the range of the available filters. This limitation poses an additional challenge in relation to the localization task, raising the "long-range interpolation problem" that has generated much interest in relation to the position coding for extended stimuli, such as Gaussian blobs and Gabor patches (Morgan and Watt, 1982; Hess and Holliday, 1992; Levi, Klein, and Wang, 1994).

## NEED FOR 3D SURFACE REPRESENTATION

One corollary of the surface reconstruction approach is a postulate that the object array is represented strictly in terms of its surfaces, as proposed by Nakayama and Shimojo (1990). Numerous studies point to a key role of surfaces in organizing the perceptual inputs into a coherent representation. Norman and Todd (1998), for example, show that discrimination of the relative depths of two visual field locations is greatly improved if the two locations to be discriminated lie in a surface rather than being presented in empty space. This result is suggestive of a surface level of interpretation, although it may simply be relying on the fact that the presence of the surface provides more information about the depth regions to be assessed. Nakayama, Shimojo, and Silverman (1989) provide many demonstrations of the importance of surfaces in perceptual organization. They show that recognition of objects (such as faces) is much enhanced where the scene interpretation allows them to form parts of a continuous surface rather than isolated pieces, even when the retinal information about the objects is identical in the two cases. This study also focuses attention on the issue of border ownership by surfaces perceived as in front of rather than behind other surfaces. Although their treatment highlights interesting issues of perceptual organization, it offers no insight into either the neural or computational mechanisms by which such structures might be achieved.

A key issue raised by the small scale of the local filters for position encoding is the mechanism of long-range interpolation. To address this question, Kontsevich and Tyler (1998) developed a sampling paradigm for object location in which the objects were defined by sampled luminance profiles in the form shown in Figure 9.3. This sampled paradigm is a powerful means for probing the properties of the luminance information contributing to shape perception. Surprisingly, the accuracy of localization by humans is almost *independent* of the sample spacing. The data showed that as few as two to three samples are all the information about a 1° Gaussian bulge that could be processed by the visual system. Adding further samples (by increasing sampling density) had no further

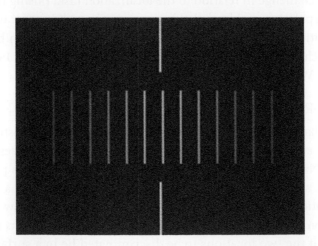

FIGURE 9.3 Sampled Gaussian luminance profile used to study long-range interpolation processes by Kontsevich and Tyler (1998; Reprinted with permisson.)

effect on the discriminability of the location of the Gaussian, as long as the intersample spacing was beyond the 3 arc min range of the local filters. Kontsevich and Tyler (1998) validated the surprisingly local prediction of this limit, finding that localization performance was invariant for spacing ranging from 30 down to 3 arc min separation, but showed a stepwise improvement (to Vernier levels) once the sample spacing came within the 3 arc min range. Sample positions were randomized to prevent them from being used as the position cue.

The implication to be drawn from this study is that some long-range interpolation mechanism is required to determine the shape of extended objects before us (because the position task is a form of shape discrimination). The ability to encode shape is degraded once the details fall outside the range of the local filters. However, the location was still specifiable to a much finer resolution than the sample spacing, implying the operation of an interpolation mechanism to determine the location of the peak of the Gaussian despite the fact that it was not consistently represented within the samples.

The conclusions from this work are that (1) the interpolation mechanism was inefficient for larger sample numbers, because it used information from only two to three samples even though up to 10 times as many samples were available; (2) the interpolation mechanism could operate over the long range to determine the shape and location of the implied object to a substantially higher precision than the spacing of the samples (~6 arc min); and (3) the mechanism was not a simple integrator over the samples within any particular range.

## EVIDENCE FOR A FUNCTIONAL ROLE OF 3D SURFACE REPRESENTATION

Likova and Tyler (2003) addressed the unitary depth map hypothesis of object localization in a 3D surface interpolation paradigm. They used sparsely sampled images adapted from those of Kontsevich and Tyler (1998) (Figure 9.4), but expanded the concept to the third dimension by incorporating a disparity cue to shape as well as the luminance cue. The luminance of the sample lines carried the luminance profile information (Figure 9.4a), and the disparity in their positions in the two eyes carried the disparity profile information (Figure 9.4b)". In this way, the two separate depth cues could be combined or segregated as needed. Both luminance and disparity profiles were identical Gaussians, and the two

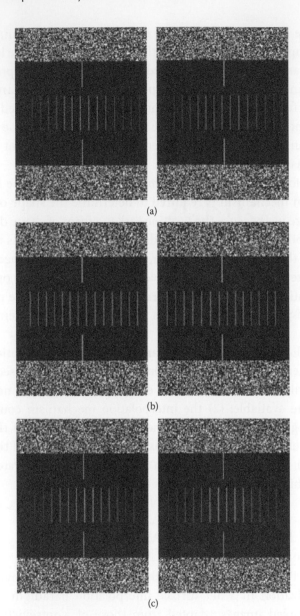

FIGURE 9.4 Stereograms showing examples of the sampled Gaussian profiles used in the Likova and Tyler (2003) experiment, defined by (a) luminance alone, (b) disparity alone, and (c) a combination of luminance and disparity. The pairs of panels should be free-fused to obtain the stereoscopic effect. (Reprinted with permission.)

types of profiles were always congruent in both peak position and width. Observers were presented with the sampled Gaussian profiles defined either by luminance modulation alone (Figure 9.4a), by disparity alone (Figure 9.4b), or by a combination of luminance and disparity defining a single Gaussian profile (Figure 9.4c). The observer's task was to make a left/right judgment on each trial of the position of the joint Gaussian bulge relative to a reference line, using whatever cues were available. Threshold performance was measured by means of the maximum-entropy Ψ staircase procedure (Kontsevich and Tyler, 1999). It should be noticeable that the luminance profile evokes a strong sense of depth as the luminance fades into the black background. If this is not evident in the printed panels, it was certainly seen clearly on the monitor screens. Free fusion of Figure 9.4b allows perception of the stereoscopic depth profile (forward for crossed fusion). Figure 9.4c shows a combination of both cues at the level that produced cancellation to flat plane under the experimental conditions. The position of local contours is unambiguous, but interpolating the peak, corresponding to reconstructing the shape of someone's nose to locate its tip, for example, is unsupportable.

Localization from disparity alone was much more accurate than from luminance alone, immediately suggesting that depth processing plays an important role in the localization of sampled stimuli (see Figure 9.5, gray circles). Localization accuracy from disparity alone was as fine as 1 to 2 arc min, requiring accurate interpolation to localize the peak of the function between the samples spaced 16 arc min apart. This performance contrasted with that for pure luminance profiles, which was about 15 arc min (Figure 9.5, horizontal line). Combining identical luminance and disparity Gaussian profiles (Figure 9.5, black circles) provides a localization performance that is qualitatively similar to that given by disparity alone (Figure 9.5, gray circles). Rather than showing the approximation to the lowest threshold for any of the functions predicted by the multiple-cue interpolation hypothesis, it again exhibits a null condition where peak localization is impossible within the range measurable in the apparatus. Contrary to the multiple-cue hypothesis, even with full luminance information, the peak position in the stimulus becomes impossible to localize as soon as it is perceived as a flat surface. This null point can only mean that luminance information, per se, is insufficient to specify the position of the luminance profile

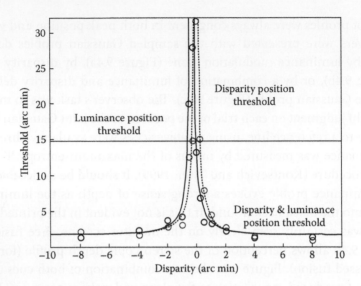

FIGURE 9.5 **(See color insert.)** Typical results of the position localization task. The gray circles are the thresholds for the profile defined only by disparity; the black circles are the thresholds defined by disparity and luminance. The dashed gray line shows the model fit for disparity alone; the solid line shows that for combined disparity and luminance, as predicted by the amount of disparity required to null the perceived depth from luminance alone (Likova and Tyler, 2003). The horizontal line shows threshold for the pure luminance. Note the leftward shift of the null point in the combined luminance/disparity function.

in this sampled stimulus. The degradation of localization accuracy can be explained only under the hypothesis that interpolation occurs within a unitary depth-cue pathway.

Perhaps the most startling aspect of the results in Figure 9.5 is that position discrimination in sampled profiles can be completely nulled by the addition of a slight disparity profile to null the perceived depth from the luminance variation. It should be emphasized that the position information from disparity was identical to the position information from luminance on each trial, so addition of the second cue would be expected to reinforce the ability to discriminate position if the two cues were processed independently. Instead, the nulling of the luminance-based position information by the depth signal implies that the luminance target is processed exclusively through the depth interpretation. Once the depth interpretation is nulled by the disparity signal, the luminance information does not support position discrimination (null point in the solid curve in Figure 9.5).

This evidence suggests that *depth surface reconstruction* is the key process in the accuracy of the localization process. It appears that visual patterns defined by different depth cues are interpreted as objects in the process of determining their location.

Evidently, the full specification of objects in general requires extensive interpolation to take place, even though some textured objects may be well defined by local information alone. The interpolated position task may therefore be regarded as more representative of real-world localization of objects than the typical Vernier acuity or other line-based localization tasks of the classic literature. It consequently seems remarkable that luminance information, per se, is unable to support localization for objects requiring interpolation. The data indicate that it is only through the interpolated depth representation that the position of the features can be recognized. One might have expected that positional localization would be a spatial form task depending on the primary form processes (Marr, 1982). The dominance of a depth representation in the performance of such tasks indicates that the depth information is not just an overlay to the 2D sketch of the positional information. Instead, it seems that a full 3D depth reconstruction of the surfaces in the scene must be completed before the position of the object is known.

A promising paradigm for studying the surface interpolation process is to use a circular version of the sampled-shape task of Likova and Tyler (2003), with targets that are Gaussian bulges defined by disparity (Figure 9.6), luminance, or motion cues. The samples may be in the form of a random-dot pattern constrained (Figure 9.6b) to a minimum dot spacing of half the average dot spacing (in all directions). Thus, the samples carrying the luminance, disparity, and motion information will always be separated by a defined distance, preventing discrimination by local filter processing. The shape/position/orientation information therefore has to be processed by some form of interpolation mechanism. Varying the luminance of the sample points of the type of Figure 9.6b in the same Gaussian shape generates the equivalent luminance figure, and such sampled figures can be used to generate the equivalent figure for structure-from-motion (cf, Regan and Hamstra, 1991, 1994; Regan, Hajdur, and Hong, 1996).

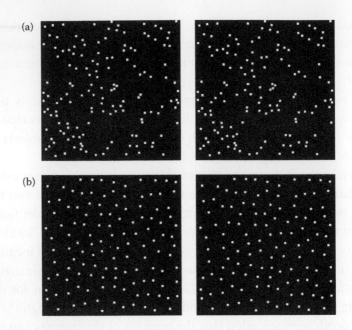

FIGURE 9.6  (a) Stereopair of a Gaussian bulge in a random-dot field. It should be viewed with crossed convergence. (b) Similar stereopair in two-dimensional sampled noise with a minimum spacing of 25 pixels. Note the perceived surface interpolation in these sampled stereopairs, and that the center of the Gaussian is not aligned with the pixel with the highest disparity.

## CORTICAL ORGANIZATION OF THE SURFACE REPRESENTATION

A striking aspect of the cortical representation of depth structure is provided by the results of a functional Magnetic Resonance Imaging (fMRI) study of the cortical responses in the human brain to disparity structure (Tyler et al., 2006). An example of the activation to static bars of disparity (presented in a dynamic noise field, with a flat disparity plane in the same dynamic noise as the null stimulus) is shown in Figure 9.7. The only patches of coherent activation (at the required statistical criterion level) are in the dorsal retinotopic areas V3A and V3B and as well as in the lateral cortex posterior to V5, in a cortical region identified as KO by the standard kinetic border localizer (Van Oostende et al., 1997), although our stimuli

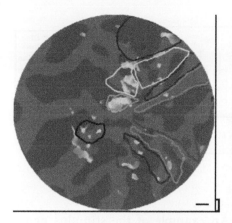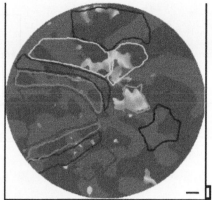

Key to retinotopic areas
V1  V2  V3  V3A/B  V4  V7  hMT+  KO

FIGURE 9.7  **(See color insert.)** Functional magnetic resonance imaging (fMRI) flat maps of the posterior pole of the two hemispheres showing the synchronized response to stereoscopic structure (yellowish phases) localized to V3A/B (yellow outlines) and KO (cyan outlines). (Reprinted with permission, from Tyler et al., 2006)

had no kinetics whatever but only static depth structure. Retinotopic analysis reveals that the dominant signal occurs at an extremely foveal eccentricity of only 0.5°.

The study of Figure 9.7 shows that one area (KO) of dorsolateral occipital cortex stands out as responding to depth structure conveyed by pure disparity cues. This result does not, however, resolve whether this cortical region is processing depth structure in general or the more specific subtype of stereoscopic depth structure in particular. In order to do so, we need a paradigm that presents the same depth structure via different depth cues. This was implemented in stimuli depicting a Gaussian bump in stereoscopic, motion, texture, and shading cues in one (test) hemifield. The subjects' task was to adjust the strength of each cue to match the depth perceived in the stereoscopic version of the stimulus in the other (comparison) hemifield. The fMRI response in the KO region for the constantly stereoscopic (comparison) and the mixed stimulus different cue

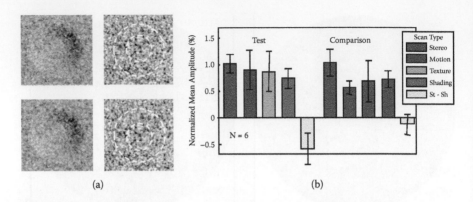

FIGURE 9.8 **(See color insert.)** (a) An example of the stimulus pairs. Gaussian bumps defined by shading (presented above and below the horizontal meridian in the left/test hemifield) and by disparity (in the right/comparison hemifield). (b) Evidence for a generic depth map in the dorsolateral occipital cortex (average of six brains). Test hemifield: Mean group cortical response to four depth cues (see key) at a dorsolateral occipital cortical location that is a candidate for the generic depth map. Note the similarity of response amplitudes for the four individual depth cues (multicolored upward bars), and no significant response in the modality alternation experiment (yellow downward bar), where disparity and shading cues are counterposed (St – Sh). Comparison hemifield: Disparity response under each condition (except the last, where the disparity was held constant as a control for the modality alternation condition). (Tyler et al., 2006)

(test) hemifields is shown in Figure 9.8. Each bar represents the activation for the contrast between full- and quarter-strength versions of the depth cue specified in the legend. The perceived depths for the four cues in the test hemifield were equated for the full-strength stimuli during each scan by performing a depth-matching task against a stereo-defined bump in the comparison hemifield.

The final condition shown in Figure 9.8 was a confrontation test for alternation between the disparity and shading cues equated for perceived depth. Thus, no response is expected in an area encoding perceived depth. This contrast cancels the signal in the response in both hemifields (yellow bars; Figure 9.8), and the slightly negative one for the purely stereoscopic hemifield may be understood as adaptation to the intensive stereoscopic stimulation. The double dissociation pattern was not seen in any other area of cortex.

## THE SURFACE CORRESPONDENCE PROBLEM

There has recently been substantial interest in the mechanism by which the motion of a surface defined purely by disparity cues is appreciated (Patterson, 1999, 2002; Lu and Sperling, 1995, 2001). For spatiotemporal structure of a stereoscopic surface, there is a fundamental ambiguity about how the surface at one time has transformed toward the surface at a later time. For example, did it move laterally, did it move in depth, or did it move in some combination of the two? This is what we term the *surface correspondence problem*, a global 4D (i.e., 3D spatiotemporal) generalization of the local correspondence problem (see Figure 9.5): what principle defines which point on the later surface corresponds to a given point on the surface at the earlier time? Much of recent computational neuroscience has been driven by two correspondence problems: (1) the *binocular correspondence problem* highlighted by Julesz (1971) and (2) its temporal counterpart of the *motion correspondence problem* delimited by the "Braddick limit" (Braddick, 1974). The temporal case generalizes to the *aperture problem* of local correspondence over time in motion analysis (Marr, 1982), with its solution via the intersection-of-constraints rule, an instantiation of the rigidity constraint. To these seminal problems we now add the *surface correspondence problem*, a meld of the stereo and motion correspondence problems, but one that operates at the next level of the processing hierarchy.

Thus, the surface correspondence problem is logically independent from, but interacts with, the two lower-level correspondence problems to achieve a global 4D solution. We expect identification of this problem to start a new series of investigations of constraints underlying its global solution to the perceived dynamics of 3D surface transformations. In particular, there is an ambiguity between the rigid lateral ($x$-axis) motion of a stereoscopically defined surface (Figure 9.9a) and the nonrigid set of local motions for the same scale surface replacement if the correspondence matching were defined by motion *orthogonal* to the surface structure (as depicted in Figure 9.9b) for sinusoidal cyclopean surfaces. Our preliminary data suggest that if a stereoscopic sinusoid is alternated between two nearby phases, we may expect to perceive it as moving laterally in a *globally rigid* solution (Figure 9.9a), despite the fact that the nearest neighbor rule would imply nonrigid shape change (Figure 9.9b; the nearest neighbor

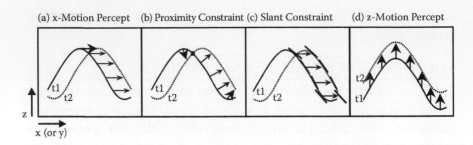

FIGURE 9.9 **(See color insert.)** (a) Diagram representing two sequential phases (full and dashed curves) of a stereoscopic sinusoidal surface (schematized as a cross-section in z,x space, i.e., a top view). Arrows show some corresponding locations, as required for the percept of lateral motion observed, while the surface waveform alternates between the two phases. (b) A proximity constraint can not explain the observed percept. (c) A surface orientation (slant) constraint would provide the requisite matches to account for the percept in (a). (d) Alternating sinusoid between near and far z-axis positions enforces a percept of z-axis stereomotion.

rule we implement as a **mutual proximity constraint** in terms of the tangent chord distance of the minimal spheres touching the two surfaces). In other words, the lateral motion manifests a higher probabilistic "weight." Further options are considered below.

This novel *surface* correspondence problem should be clearly distinguished from the issue of the *token* correspondence problem required to construct surfaces from motion cues (Grimson, 1982; Green, 1986; He and Nakayama, 1994). Such studies are concerned with the correspondence between local luminance elements making up the motion, which then defines the surface structure by means of differential *motion* cues—the classic structure-from-motion paradigm. In the present case, the surface is defined stereoscopically and then alternates between two depth configurations. The question is, what defines the correspondence between the surface locations at $t_1$ and $t_2$? This question is independent of the elements defining the surface (such as the dots shown in Figure 9.6a): the elements can completely change between $t_1$ and $t_2$ without disrupting the surface correspondence. For example, if there is no change in the *surface* configuration, no surface motion will be seen even if the dynamic noise elements completely change their location and character between $t_1$ and $t_2$.

In terms of psychophysics, the surface correspondence problem is quantitative rather than qualitative, because there is bound to be a balance point in the jump size where the motion shifts from lateral $(x,y)$ to depth $(z)$ motion of the stereoscopic surface, or stereomotion. However, the interpretation of this transition is that the local correspondence naturally leads to a percept of lateral stereomotion, because the local disparity is changing. In order to see lateral motion of the cyclopean surface, the system must be performing a surface-based correspondence match, as illustrated in Figure 9.9c, for a surface jumping between a $t_1$ configuration and a $t_2$ configuration.

All the closest points in any Euclidean metric in 3D space (regardless of the relative horizontal/vertical scaling) must be reassigned in order to generate a uniform lateral motion. This reassignment may be regarded as the valid signature of some global process operating to generate the lateral motion percept of the depth surface. For example, a surface-based match would be selective for surface slant and, hence, could disambiguate the matches by matching only to neighbors of the same surface slant (Figure 9.9c). Thus, a surface-based match would enforce lateral cyclopean motion because the nearest patch of surface with the same slant is always in the pure lateral direction (for a purely lateral displacement of the computed surface). In particular, the surface slant constraint would interdict the stability of any intersection point between the surfaces because the surfaces have opposite slants at the intersection points.

The evidence reviewed in this overview points toward the key role of the surface representation in providing the "glue" or "shrink-wrap" to link the object components in their appropriate relationships. It also emphasizes the inherent three-dimensionality of this surface shrink-wrap.

The fact that the replacement of sinusoidal disparity structure is perceived as motion of that structure (see Figure 9.9a) raises the question of whether a stereoscopic structure moving in 3D space is processed by the same cortical mechanism as static disparity structure, and whether the z-axis motion of such structure is processed by the same mechanism as its $y$-axis motion. To resolve this question, we needed a paradigm to dissociate responses to a 3D structure when static, laterally moving ($y$-axis motion) or moving in depth ($z$-axis motion). If the corresponding three stimuli activated the same cortical site, that would not help with the answer, because such a site could be the basis for a variety of inferences about the neural substrate for the depth structure. But if any two stimuli activate. But if any two stimuli activate different cortical sites, it provides strong evidence of a difference in the underlying neural mechanisms.

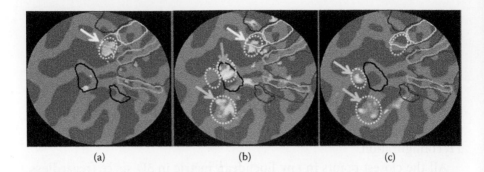

(a)                    (b)                    (c)

FIGURE 9.10  **(See color insert.)** Occipital flat map for the left hemisphere of one subject shows distinct locations of significant activation (yellowish patches). Full-colored outlines show retinotopic areas as in Figure 9.7. Dark blue outline: boundary of hMT+ defined by a motion localizer. Dashed outlines are for comparison of clusters of activated voxels across the three conditions. (a) Stereoscopic structure of a static sinusoidal disparity versus a flat disparity-plane activates a region in the dorsolateral cortex. (b) Frontoparallel ($y$-axis) stereomotion of the sinusoidal stereoscopic surface contrasted with a flat plane activates a swath of cortex including hMT+ (green arrow), together with two sites from the previous two conditions: the depth-structure region similar to (a) (white arrow) and the ventral site seen in (c) (cyan arrow). (c) $Z$-axis stereomotion versus $y$-axis stereomotion of the same stereoscopic sinusoidal surface activates regions anterior (yellow arrow; cyclopean stereomotion area CSM) and ventral (cyan arrow) to hMT+.

In fact, robust differential activations are observed in cortical areas in lateral cortex, as illustrated in Figure 9.10. The interpretations are that both $y$ and $z$ stereomotion are processed separately from depth structure, per se; that frontoparallel stereomotion is processed similarly to lateral luminance-based motion in hMT+ but also activates a second ventral site; and that $z$-axis stereomotion is encoded at a different level in the processing hierarchy (Likova and Tyler, 2007) that also includes the ventral site.

The implications of the processing sequence for surface interpolation and stereomotion may be captured in a flow diagram (Figure 9.11). Local disparity is known to be processed in V1, although not probed by the experiments of Figure 9.10, because local disparity activation was equated in all three comparisons. The disparity signals must reach the dorsal region for *depth interpolation* of the surface structure in the static disparity image.

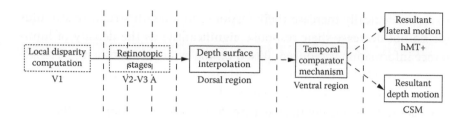

FIGURE 9.11 Flow diagram of the processing implied by the activations in Figure 9.10. Boxes show processing stages with labels indicating their cortical sites. Arrows represent connecting pathways, dashed when speculative. Vertical dashed lines indicate logical sequence separators based on prior studies.

The ventral region is activated only by temporal changes in the surface structure and is therefore likely to be a *temporal comparator mechanism* operating for both types of depth motion. Such a mechanism would need input from a surface interpolation mechanism, but we do not have direct evidence of a pathway connecting the dorsal and ventral regions.

The lack of dorsal activation in Figure 9.10c implies not that there was no activation by surface structure but that there was no change in the surface structure to generate activation as the motion direction changed from z-axis (test) to y-axis (control); in contrast, the conditions for Figure 9.10a,b involved changes in depth structure, as the control was a flat stereosurface. The final element of the flow diagram of Figure 9.11 is a split into separate representations for the y-axis (hMT+ region) and z-axis (CSM region) directions of cyclopean motion. The separation is mandated by the fMRI activation, but the connections remain speculative (as did those of Maunsell and Van Essen, 1983, in their early neurophysiological studies). Further manipulations will be required to resolve all the details of the flow diagram.

## CONCLUSION

The concept of surface representation requires a surface interpolation mechanism to represent the surface in regions of the visual field where the information is undefined. Whereas a typical receptive-field summation mechanism shows a stronger response as the amount of stimulus information increases, the characteristic of an interpolation mechanism is to increase its response as stimulus information is *reduced* and more extended interpolation is required. The cortical locus of the interpolation mechanism may therefore be sought by identifying fMRI activation sites that paradoxically increase their response to a depth structure stimulus (or do not decrease their response significantly) as the density of luminance information is reduced. Based on the evidence that there is a single depth interpolation mechanism for all visual modalities (Tyler et al., 2006), experiments should be conducted for disparity, motion, and texture density cues to the same depth structure, to test the prediction that the same cortical interpolation site will show increased response as dot density is decreased for all three depth structure cues.

In conclusion, the primary outcome of this review is the concept of *3D surface interpolation*, that is, that the predominant mode of spatial processing is through a self-organizing surface representation (or *attentional shroud*, see Ch. 0) within a full 3D spatial metric. It is not until such a surface representation is developed that the perceptual system seems to be able to localize the components of the scene (Likova & Tyler, 2003). This view is radically opposed to the more conventional concept that the primary quality of visual stimuli is their location, with other properties attached to this location coordinate (Marr, 1982). By contrast, the concept of the attentional shroud is a flexible network for the internal representation of the external object structure. In this concept, the attentional shroud is, itself, the perceptual coordinate frame. It organizes ("shrink-wraps") itself to optimize the spatial interpretation implied by the complex of binocular and monocular depth cues derived from the retinal images. It is not until this depth reconstruction process is complete that the coordinate locations can be assigned to the external scene. In this sense, localization is *secondary* to the full depth representation of the visual input. Spatial form, usually seen as a predominantly 2D property that can be rotated into the third dimension, becomes a primary 3D concept of which the 2D projection is a derivative feature.

The net result of this analysis is to offer a novel insight into the nature of the binding problem. The separate stimulus properties and local features are bound into a coherent object by the "glue" of the *global 3D surface* representation. This view is a radical counterpoint to the concept of breaking the scene down into its component elements by means of specialized receptive fields and recognition circuitry. The evidence reviewed in this overview emphasizes the *inherent three-dimensionality* of the surface "shrink-wrapping" process by the attentional shroud in the form of a prehensile matrix that can cohere the object components whose images are projected onto the sensorium. Such *active binding* processes are readily implementable computationally with plausible neural components that could reside in a locus of 3D reconstruction in the human cortex.

# CHAPTER 10

# 3D and Spatiotemporal Interpolation in Object and Surface Formation

Philip J. Kellman

Patrick Garrigan

Evan M. Palmer

## CONTENTS

## INTRODUCTION

As David Marr observed in his classic book *Vision* (1982), understanding visual perception involves issues at multiple levels of analysis. This observation applies not only to conceptual differences in the kinds of questions researchers must ask, but also to different levels of visual processing. Vision researchers have made great progress in understanding early cortical filtering. At the opposite end, research has revealed some areas in which high-level representations reside, such as those for objects or faces. Between these levels, however, there is a considerable gap. This gap in "middle vision" involves all of Marr's levels: the understanding of information for computing representations of contours, surfaces, and objects; the representations and processes involved; and the sites and roles of cortical areas. Although Marr emphasized that these levels have substantial independence in terms of the questions they pose, the lag in understanding what goes on "in the middle" is also related to interactions among these levels. Understanding the task and information paves the way for process descriptions. Similarly, detailed hypotheses about processes and representations guide meaningful neurophysiological investigations.

Fundamental to the middle game in vision are three-dimensional (3D) representations. What is the shape of a surface? How do we represent the shapes of 3D objects and obtain these representations from incomplete and fragmentary projections of an object to the eyes? How do we obtain descriptions of objects and surfaces in ordinary environments, where the views of most objects are partly obstructed by other objects, and visible areas change in complex ways as objects and observers move?

Although many traditional approaches to vision have sought to discover how meaningful perceptual representations can be gotten by inferences from static, two-dimensional (2D) images, it has become increasingly clear that human vision both utilizes complex 3D and spatiotemporal information as inputs and constructs 3D surface representations as outputs. Although human vision may exploit shortcuts for some tasks, 3D surface representations play many important roles both in our comprehension of the world and our ability to interact with it.

In this chapter, we consider several lines of research aimed at improving our understanding of 3D and spatiotemporal surface and object formation. Specifically, we are concerned with the achievement of surface or object representations when the visual system must interpolate across spatial and spatiotemporal gaps in the input. The human visual system possesses remarkable mechanisms for recovering coherent objects and surface representations from fragmentary input. Specifically, object and surface perception depend on interpolation processes that overcome gaps in contours and surfaces in 2D, 3D, and spatiotemporal displays. Recent research suggests that the mechanisms for doing so are deeply related in that they exploit common geometric regularities.

## SOME PHENOMENA OF VISUAL INTERPOLATION

In ordinary perception, partial occlusion of objects and surfaces is pervasive. Panels (a)–(d) on the left side of Figure 10.1, for example, show views of a house occluded by a fence. Even in a single static view, we are able to get some representation of the scene behind the fence. If the several views were seen in sequence by a walking observer, we would get a remarkably complete representation, as suggested by Figure 10.1c.

Perceiving whole objects and continuous surfaces requires perceptual processes that connect visible regions across gaps in the input to achieve accurate representations of unity and shape. These have most often been studied for static 2D representations. Yet perception grapples with a 3D world and produces, in part, truly 3D representations of object contours and surfaces. Furthermore, when objects or observers move, the visible regions of objects change over time, complicating the requirements of object formation. The system deals with fragmentation, not only in space, but across time as well. Thus, we may think of contour and surface perception in the real world as a mapping from information arrayed across four dimensions (three spatial dimensions and time) into 3D spatial representations. If motion is represented, visual object and surface formation is a mapping from fragmented four-dimensional (4D) inputs into coherent, functionally meaningful, 4D representations.

These phenomena are formally similar in that the same physically specified contours of the central figure are given in Figure 10a, 10b, and 10c, and the completed object in each case is defined by the same collection of physically specified and interpolated contours. (Figure 10d includes only the corresponding interpolations in the middle part of the figure.)

### Categories of Interpolation Phenomena

A number of phenomena involve connecting visible contours and surfaces across gaps (Figure 10.2). Figure 10.2a shows partial occlusion. Six

FIGURE 10.1 Real-world interpolation requires integration over time and space. Frames (a,b,c,d): Several images of an occluded real-world scene. The porch of this house is visible between the fence posts and is perceived as a series of connected visual units despite the fact that shape information is fragmented in the retinal projection. Frame (e): When motion and three-dimensional contour and surface interpolation operate, the visual system can generate a far more complete representation of the scene, of the sort depicted here.

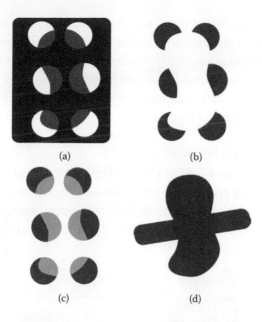

FIGURE 10.2 **(See color insert.)** Four perceptual phenomena that can be explained by the same contour interpolation process. (a) A partially occluded object. The blue fragments are spatially disconnected, but we perceive them as part of the same object. (b) The same shape appears as an illusory figure and is defined by six circles with regions removed. (c) A bistable figure that can appear either as a transparent blue surface in front of six circles or an opaque blue surface seen through six circular windows. (d) A self-splitting object. The homogenous black region is divided into two shapes. This figure is bistable because the two shapes appear to reverse depth ordering over time.

noncontiguous blue regions appear; yet, your visual system connects them into a single object extending behind the black occluder. The object's overall shape is apparent. Perceptual organization of this scene also leads to the perception of circular apertures in the black surface, through which the blue object and a more distant white surface are seen. Figure 10.2b illustrates the related phenomenon of *illusory contours* or *illusory objects*. Here, the visual system connects contours across gaps to create the central white figure that appears in front of other surfaces in the array. Figure 10.2c shows a transparency version of an illusory figure; the figure is created but one can also see through it. Finally, in Figure 10.2d, a uniform black region is seen to split into two visible figures, a phenomenon that has been called *self-splitting objects*.

These phenomena are formally similar in that the same physically specified contours of the central figure are given in each case, and the completed object in each case is defined by the same collection of physically specified and interpolated contours.

### Contour and Surface Processes

Evidence suggests two kinds of mechanisms for connecting visible areas across gaps: contour and surface interpolation. These processes can be distinguished because they operate in different circumstances and depend on different variables (See Figure 10.3). Contour interpolation depends

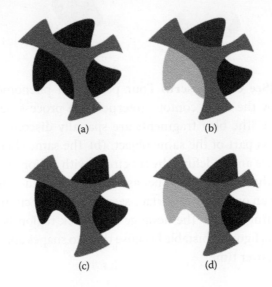

FIGURE 10.3 **(See color insert.)** Contour and surface interpolation. (a) The three black regions appear as one object behind the gray occluder. Both contour and surface interpolation processes are engaged by this display. (b) Contour interpolation alone. By changing the surface colors of visible regions, surface interpolation is blocked. However, the relations of contours still engage contour interpolation, leading to some perceived unity of the object. (c) Surface interpolation alone. By disrupting contour relatability, contour interpolation is blocked. Due to surface interpolation, there is still some impression that the three fragments connect behind the occluder. (d) With both contour and surface interpolation disrupted, blue, yellow, and black regions appear as three separate objects.

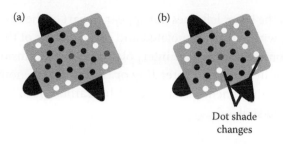

(a)　　　　　　(b)

Dot shade
changes

FIGURE 10.4 **(See color insert.)** Illustration of two- dimensional surface interpolation. The circular areas in the display do not trigger contour processes, due to the absence of tangent discontinuities. Surface interpolation causes some circular areas to appear as holes in the occluder rather than as spots in front. Two dots in (a) are changed in color in (b), causing a difference in their appearance (e.g., the yellow spot in (a) when turned white becomes a hole due to its relation with the color of the surround). Relations of contour and surface interpolation are shown by blue spots appearing as holes if they fall within interpolated (or extrapolated) contours of the blue display.

on geometric relations of visible contour segments that lead into contour junctions. Surface interpolation in 2D displays can occur in the absence of contour segments or junctions; it depends on the similarity of lightness, color, or texture of visible surface patches.

Figure 10.4 illustrates the action of surface interpolation. Some of the circles in the display, such as the yellow ones, appear as spots on the surface. In contrast, most of the blue circles appear to be part of a single, occluded, blue figure, visible through holes. The white spots also appear to be holes rather than spots; through them, the white background surface is seen. These perceptual experiences arise from the surface interpolation process. Visible regions are connected across gaps in the input based on the similarity of their surface qualities (e.g., lightness, color, and texture). These connections cannot be given by contour interpolation, as the circles have no contour junctions. Certain rules govern surface interpolation; for example, it is confined by real and interpolated edges (Yin, Kellman, and Shipley, 2000). In the figure, note that the rightmost circle does not link up with the occluded object. This result occurs because that dot does not fall within real or interpolated contours of the blue object. Whereas contour interpolation processes are relatively insensitive to relations of lightness or color, the surface process depends crucially on these. Notice that the yellow dot on the lower left does *not* appear as part of the occluded object, despite being within the interpolated and real contours of the blue object.

This phenomenon of surface interpolation under occlusion appears to be one of a family of surface spreading or "filling-in" phenomena, such as the color-spreading phenomena studied by Yarbus (1967) and filling-in across the blind spot.

## A MODEL OF CONTOUR INTERPOLATION IN STATIC 2D SCENES

Complementary processes of contour and surface interpolation work in concert to connect object fragments across gaps in the retinal image and recover the shape of occluded objects (e.g., Grossberg and Mingolla, 1985; Kellman and Shipley, 1991). Interpolated boundaries of objects, whether occluded or illusory, constrain spreading of surfaces across unspecified regions in the image, even if the interpolated boundaries are not connected to others (Yin, Kellman, and Shipley, 1997, 2000). Here, we briefly review the context for developing a 4D model of contour interpolation and surface perception.

### The Geometry of Visual Interpolation

A primary question in understanding visual object and surface formation is what stimulus relationships cause it to occur? Answering this question is fundamental in several respects. It allows us to understand the nature of visual interpolation. Some visible fragments get connected, whereas others do not. Discovering the geometric relations and related stimulus conditions that lead to object formation is analogous to understanding the grammar of a language (e.g., what constitutes a well-formed sentence). Understanding at this level is also crucial for appreciating the deepest links between the physical world and our mental representations of it. Characterizing the stimulus relations leading to object formation is at first descriptive, but as unifying principles are revealed, they help us to relate the information used by the visual system to the physical laws governing the projection of surfaces to the eyes, whether these are deep constraints about the way the world works (e.g., Gibson, 1979; Marr, 1982) or scene statistics (e.g., Geisler et al., 2001).

### Initiating Conditions for Interpolation

An important fact about contour interpolation is that the locations of interpolated contours are highly restricted in visual scenes. In general, interpolated contours begin and end at junctions or corners in visible contours (tangent discontinuities)—locations at which contours have no unique orientation (Shipley and Kellman, 1990; Rubin, 2001). Some have suggested that second-order discontinuities (points that are first-order continuous but mark a change in curvature) might also weakly trigger interpolation (Shipley and Kellman, 1990; Albert and Hoffman, 2000; Albert and Tse, 2000; Albert, 2001; for recent discussion see Kellman, Garrigan, and Shipley, 2005). Tangent discontinuities arise from the optics of how occluded objects project to the eyes: it can be proven that the optical projection of one object occluding another will contain these image features (Kellman & Shipley, 1991). Shipley and Kellman (1990) observed that, in general, interpolated contours begin and end at tangent discontinuities and showed that their removal eliminated or markedly reduced contour interpolation. Heitger et al. (1992) called tangent discontinuities "key points" and proposed a neurally plausible model for their extraction from images. The presence or absence of tangent discontinuities can be manipulated in illusory contour images by rounding the corners of inducing elements, which weakens contour interpolation (e.g., Albert and Hoffman, 2000; Kellman et al., 2005; Shipley and Kellman, 1990; Palmer, Kellman, and Shipley, 2006).

## Contour Relatability

What determines which visible contour fragments get connected to form objects? Although tangent discontinuities are ordinarily necessary conditions for contour interpolation, they are not sufficient. After all, many corners in images are corners of objects, not points at which some contour passes behind an intervening surface (or in front, as in illusory contours).

Contour interpolation depends crucially on geometric relations of visible contour fragments, specifically the relative positions and orientations of pairs of edges leading into points of tangent discontinuity. These relations have been described formally in terms of *contour relatability* (Kellman and Shipley, 1991; Singh and Hoffman, 1999a). Relatability is a mathematical notion that defines a categorical distinction between edges that can connect by interpolation and those that cannot (see Kellman and Shipley, 1991, 175–177). The key idea in contour relatability is smoothness (e.g., interpolated contours are differentiable at least once), but it also incorporates monotonicity (interpolated contours

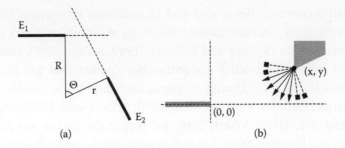

(a)　　　　　　　　　　　　　(b)

FIGURE 10.5　Contour relatability describes formally a categorical distinction between edges that can be connected by visual interpolation and those that cannot. (a) Geometric construction defining contour relatability (see text). (b) Alternative expression of relatability. Given one visible contour fragment terminating in a contour junction at (0,0) and having orientation 0°, those orientations $\theta$ that satisfy the equation $\tan^{-1}(y/x) \leq \theta \leq \Phi/2$ are relatable. In the diagram, these are shown with solid lines, whereas nonrelatable orientations are shown with dotted lines.

bend in only one direction), and a 90° limit (interpolated contours bend through no more than 90°). Figure 10.5 shows a construction that is useful in defining contour relatability. Formally, if $E_1$ and $E_2$ are surface edges, and $R$ and $r$ are perpendicular to these edges at points of tangent discontinuity, then $E_1$ and $E_2$ are relatable if and only if

$$0 \leq R\cos\theta \leq r \qquad (10.1)$$

Although the precise shape of interpolated contours is a matter of some disagreement, there are two properties of relatability that cohere naturally with a particular class of contour shapes. First, it can be shown that interpolated edges meeting the relatability criteria can always be comprised of one constant curvature segment and one zero curvature segment. Second, it appears that this shape of interpolated edges has the property of being a minimum curvature solution in that it has lowest maximum curvature: any other first-order continuous curve will have at least one point of greater curvature (see Skeath, 1991, in Kellman and Shipley, 1991). This is a slightly different minimum curvature notion than minimum energy.

## One Object or Two?

Relatability defines a categorical distinction—which relative positions and orientations allow edges to be connected by contour interpolation. Such a

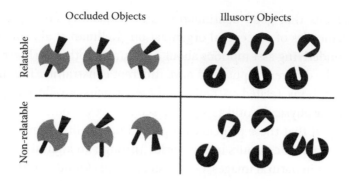

FIGURE 10.6    Examples of relatable and nonrelatable contours.

distinction is important, as object perception often involves a discrete determination of whether two visible fragments are part of the same object or not. Figure 10.6 shows examples of relatable and nonrelatable edges, in both perception of partly occluded objects and perception of illusory objects. Complete objects are formed in the top row but not in the bottom row. Object formation has profound effects on further processing, such as generation of a representation of missing areas, generation of an overall shape description, and comparison with items or categories in memory. Research indicates that the representation of visual areas as part of a single object or different objects has many important effects on information processing (Baylis and Driver, 1993; Zemel et al., 2002; Kellman, Garrigan, and Shipley, 2005).

## Quantitative Variation

Although the discrete classification of visible areas as connected or separate is important, there is also reason to believe that quantitative variation exists within the category of relatable edges (Kellman and Shipley, 1991; Banton and Levi, 1992; Shipley and Kellman, 1992a, 1992b; Field, Hayes, and Hess, 1993; Singh and Hoffman, 1999b). For example, experiments indicating a decline to a limit around 90° were reported by Field, Hayes, and Hess (1993). Singh and Hoffman (1999a) proposed an expression for quantitative decline of relatability with angular change.

## Ecological Foundations

The notion of relatability is sometimes described as a formalization of the Gestalt principle of good continuation (Wertheimer, 1923/1938). Recent work suggests that good continuation and relatability are separate but related principles of perceptual organization (Kellman et al., 2003). Both embody underlying assumptions about contour smoothness (Marr, 1982), but they take different inputs and have different constraints. The smoothness assumptions related to both of these principles reflect important aspects of the physical world as it projects to the eyes. Studies of image statistics suggest that these principles approach optimality in matching the structure of actual contours in the world. Through an analysis of contour relationships in natural images, Geisler et al. (2001) found that the statistical regularities governing the probability of two edge elements cooccurring correlate highly with the geometry of relatability. Two visible edge segments associated with the same contour meet the mathematical relatability criterion far more often than not.

## 3D CONTOUR INTERPOLATION

Object formation processes are three-dimensional. Figure 10.7 gives an example—a stereogram that may be free-fused by crossing the eyes. One sees a vivid transparent surface with a definite 3D shape. Object formation takes as inputs 3D positions and orientations of edges and produces as outputs 3D structures (Kellman and Shipley, 1991; Carman and Welch, 1992; Kellman et al., 2005; Kellman, Garrigan, and Shipley, 2005).

Until recently, there has been no account of the stimulus conditions that produce 3D interpolation. Kellman et al. (2005) proposed that 3D interpolation might be governed by a straightforward 3D generalization of

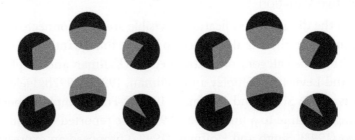

FIGURE 10.7 **(See color insert.)** Three-dimensional (3D) interpolation. The display is a stereogram that may be free-fused by crossing the eyes. Specification of input edges' positions and orientations in 3D space (here given by stereoscopic disparity) leads to creation of a vivid, connected, transparent surface bending in depth.

2D relatability. As in the 2D case, interpolated contours between 3D edges must be smooth, monotonic, and bend no more than 90°. Similarly, where 3D interpolated contours meet physically given edges, the orientations of the physically given part and the interpolated part must match.

Formally, we define, for a given edge and any arbitrary point, the range of orientations that fall within the limits of relatability at that point. In the Cartesian coordinate system, let $\Theta$ be an angle in the $x$-$y$ plane, and $\varphi$ an angle in the $x$-$z$ plane (for simplicity, in both cases zero degrees is the orientation parallel to the $x$-axis). Positioning one edge with orientation $\Theta = \varphi = 0$ and ending at the point $(0, 0, 0)$, and positioning a second edge at $(x,y,z)$ somewhere in the volume with $x > 0$, the range of possible orientations $(\theta, \varphi)$ for 3D-relatable edges terminating at that point are given by

$$\tan^{-1}\left(\frac{y}{x}\right) \leq \theta \leq \frac{\pi}{2} \tag{10.2}$$

and

$$\tan^{-1}\left(\frac{z}{x}\right) \leq \varphi \leq \frac{\pi}{2} \tag{10.3}$$

As in the 2D case, we would expect quantitative variation in the strength of interpolation within these limits. The lower bounds of these equations express the absolute orientation difference (180° for two collinear edges ending in opposite directions) between the reference edge (edge at the origin) and an edge ending at the arbitrary point oriented so that its linear extension intersects the tip of the reference edge. The upper bounds incorporate the 90° constraint in three dimensions.

How might the categorical limits implied by the formal definition of 3D relatability be realized in neural architecture? In 2D cases, it has been suggested that interpolation occurs through lateral connections among contrast-sensitive oriented units having particular relations (Field, Hayes, and Hess, 1993; Yen and Finkel, 1998). Analogously, 3D relatability specifies a "relatability field" or volume within which relatable contour edges can be located. At every location in the volume, relatable contours must have a 3D orientation within a particular range specific to that location.

The interpolation field suggests that, contradictory to some 2D models of contour interpolation, early visual cortical areas that do not explicitly code 3D positions and contour orientations may be insufficient for the neural implementation of 3D contour interpolation (Kellman, Garrigan, and

Shipley, 2005). As we discuss below, there are interesting considerations regarding exactly where the neural locus of contour interpolation may be.

## Experimental Studies of Contour Interpolation

An objective performance paradigm for testing 3D contour relatability was devised by Kellman, Garrigan, Shipley, Yin, and Machado (2005) and is illustrated in Figure 10.8. In their experiments, subjects were shown stereoscopically presented 3D planes whose edges were either relatable or not. Examples of relatable and nonrelatable pairs of planes are shown in the columns of Figure 10.8. Orthogonal to relatability are two classes of stimuli, converging and parallel planes, shown in the rows of Figure 10.8. In these experiments, subjects were asked to classify stimuli like the ones shown as either parallel or converging. The idea is that, to the extent that 3D relatability leads to object formation, judging the relative orientations of 3D relatable planes should be easier than judging the relative orientations of 3D nonrelatable planes.

Kellman et al. (2005) found that subjects could make this classification more accurately and quickly when the planes were 3D relatable. This result

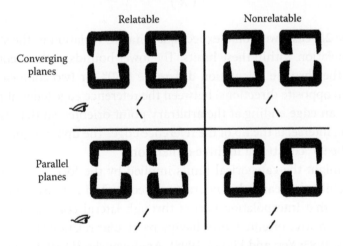

FIGURE 10.8 Experimental stimuli used to test three-dimensional (3D) object formation from 3D relatability. It was predicted that sensitivity and speed in classifying displays like these as either converging or parallel would be superior for displays in which unitary objects were formed across the gaps by contour interpolation, and that object formation would be constrained by 3D relatability. Both predictions were confirmed experimentally (Kellman et al., 2005).

is consistent with 3D relatability as a description of the geometric limits of 3D contour interpolation and object formation. A variety of other experiments indicated that the results depended on 3D interpolation, rather than some other variable, such as an advantage of certain geometric positions for making slant comparisons. (For details, see Kellman et al., 2005.)

## 3D SURFACE INTERPOLATION

Contour and surface processes often work in complementary fashion (Grossberg and Mingolla, 1985; Nakayama, Shimojo, and Silverman, 1989; Yin, Kellman, and Shipley, 1997, 2000; Kellman, Garrigan, and Shipley, 2005). Studies with 2D displays have shown that surface interpolation alone can link areas under occlusion based on similarity of surface quality. Surface similarity may be especially important in 2D, because all visible surface regions are confined to the same plane. In 3D, the situation is different. Here, geometric positions and orientations of visible surface patches may also be relevant.

We have recently been studying whether 3D surface interpolation depends on geometric constraints and, if so, how these relate to the constraints that determine contour interpolation. To study 3D surface interpolation apart from contour processes, we use visible surface patches that have no oriented edges. These are viewed through apertures (Figure 10.9).

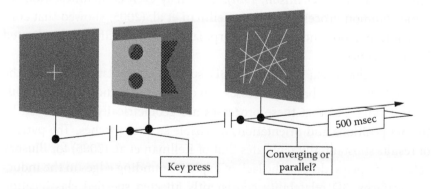

FIGURE 10.9 Use of the parallel/converging method for studying three-dimensional (3D) surface interpolation. A fixation point is followed by a display in which surface patches slanted in depth are viewed through two apertures. Participants make a forced choice as to whether the visible surface patches were in parallel or converging planes. (From Fantoni et al., 2008. With permission.)

We used a version of the parallel/converging method to study 3D surface interpolation. Displays were made of dot-texture surfaces; due to their lack of oriented edges, these surface patches could not support contour interpolation. Participants made a forced choice on each trial as to whether two surface patches, visible through apertures, lay in parallel or converging (intersecting) planes. As in 3D contour interpolation, we hypothesized that completion of a connected surface behind the occluder would facilitate accuracy and speed on this task. We also tested whether 3D relatability—applied to the orientations of surface patches rather than contours—might determine which patches were seen, and processed, as connected. 3D relatable patches were compared to displays in which one patch or the other was shifted to disrupt 3D relatability.

Figure 10.10 shows representative data on 3D surface interpolation (Fantoni et al., 2008). As predicted, 3D relatable surface patches showed sensitivity and speed advantages over nonrelatable surface patches. This effect was just as strong for vertically misaligned apertures as for vertically aligned ones. Consistent with a 90° constraint, the difference between 3D relatable and nonrelatable conditions decreased as the slant of each patch approached 45° (making their relative angle approach 90°). Many questions remain to be investigated, but these results suggest the fascinating possibility that both contour and surface interpolation in 3D share a common geometry (cf, Grimson, 1981). They may even be manifestations of some common process, although Kellman et al. (2005) showed that contour interpolation, not surface interpolation, was specifically implicated in their results.

The results of experiments on 3D surface interpolation support the notion that surface-based processes can operate independently of contour information, and that these processes are geometrically constrained by the 3D positions and orientations of visible surface patches. The pattern of results substantially replicates that of Kellman et al. (2005) for illusory contour displays, despite the lack of explicit bounding edges in the inducing surfaces. 3D relatability consistently affected speeded classification performance, by facilitating it for 3D relatable displays relative to displays in which 3D relatability was disrupted by both a depth shift of one surface relative to the other (violating the monotonicity constraint) and large values of relative stereo slant (violating the 90° constraint in converging displays).

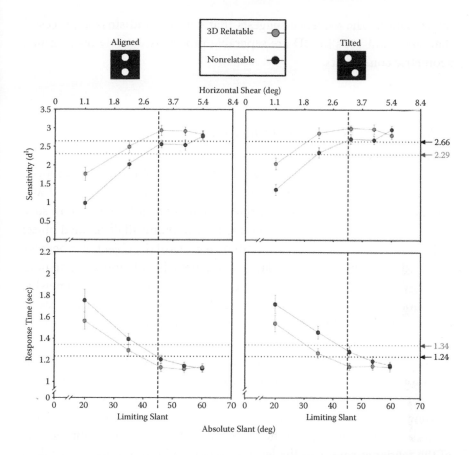

FIGURE 10.10  **(See color insert.)** Three dimensional (3D) surface interpolation data. Sensitivity (upper panels) and response times (lower panels) for 3D relatable and 3D nonrelatable surface patches in aligned (right) and misaligned (left) aperture configurations. (From Fantoni et al., 2008. With permission.)

2D surface interpolation may constitute a special case of a more general 3D process. In 3D, the primary determinant of interpolation may be geometric relations, not similarity of surface quality. In our displays, position and orientation of surface patches seen through apertures were specified by binocular disparity, along with information from vergence. It appears that disparity provided sufficient information for the extraction of the 3D orientation of inducing patches necessary to constrain surface interpolation. The evidence suggests that contour and surface processes that surmount gaps in 3D are separable processes but rely on common geometric constraints.

## SPATIOTEMPORAL INTERPOLATION

Despite the fact that perception in the laboratory is often studied with well-controlled, static, 2D images, ordinary perception usually involves diverse, moving, 3D objects. When an object is partially occluded and moves relative to the occluding surface, it is *dynamically occluded*. In such circumstances, shape information from the dynamically occluded object is discontinuous in both space and time. Regions of the object may become visible at different times and places in the visual field, and some regions may never project to the observer's eyes at all. Such cases are somewhat analogous to static, 2D occluded images, except that the partner edges on either side of an occluding boundary may appear at different times and be spatially misaligned.

For instance, imagine standing in a park and looking past a grove of trees toward a street in the distance. A car drives down the street from left to right and is visible beyond the grove. The car goes into and out of view as it passes behind the tree trunks and tiny bits and pieces of it twinkle through the gaps in the branches and leaves. You might see a bit of the fender at time 1 in the left visual field, a bit of the passenger door at time 2 in the middle of the visual field, and a bit of the trunk at time 3 in the right visual field. But what you perceive is not a collection of car bits flickering into and out of view. What you perceive is a *car*, whole and unified. In other words, your visual system naturally takes into account the constantly changing stimulation from a dynamically occluded object, collects it over time, compensates for its lateral displacement, and delivers a coherent percept of a whole object.

This feat of perception is rather amazing. Given that boundary inter-polation for static images declines as a function of spatial misalign-ment, and given that the pieces of the car became visible at different places throughout the visual field, our perception of a coherent object is quite remarkable. The key is that the spatially misaligned pieces did not appear at the same time, but rather in an orderly temporal progression as the car moved. What unifies the spatial and temporal elements of this equation is, of course, motion. The motion vector of the car allows the visual system to correct for and anticipate the introduction of new shape information.

Palmer, Kellman, and Shipley (2006) considered the requirements for spatiotemporal object formation. One important requirement is *persistence*: In order to be connected with fragments not yet visible, a momentarily viewed fragment must be represented after it becomes occluded. (See Figure 10.11 at time $t_0$.) A second requirement is *position updating*. Not only must a previously viewed, moving fragment continue to be represented, its spatial position must be updated over time, in accordance with its previously observed velocity (Figure 10.11 at time $t_1$). Finally, previously viewed and currently viewed fragments are both utilized by processes of contour and surface interpolation, which connect regions across spatial gaps (Figure 10.11 at $t_2$). When new fragments of the object come into view, they are integrated with the persisting, position-updated

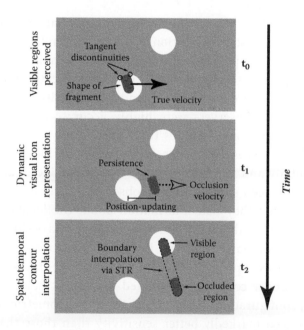

FIGURE 10.11 Spatiotemporal interpolation processes. An object moves from left to right behind an occluder with two circular windows. At $t_0$, shape and motion information about visible regions of the rod are perceived. At $t_1$, the shape and current position of the occluded region of the rod are represented in the dynamic visual icon representation. At $t_2$, another portion of the rod becomes visible, and contour interpolation occurs between the occluded and visible regions.

fragments via contour and surface interpolation processes that have previously been identified for static objects. As a result of persistence and updating processes, whether contours are interpolated is constrained by the same geometric relations of contour relatability (Kellman and Shipley, 1991; Kellman, Garrigan, and Shipley, 2005) that determine unit formation in static arrays.

As the spatial relationships that support interpolation are highly constrained, accurate representations of previously seen fragments are important for allowing spatiotemporal object formation to occur, and to operate accurately. Palmer, Kellman, and Shipley (2006) combined the requirements for object formation with proposals about visual mechanisms that represent, update, and connect object fragments over time in a model of spatiotemporal relatability (STR). The model provides an account for perception of dynamically occluded objects in situations such as that presented in Figure 10.1.

In a series of experiments, Palmer, Kellman, and Shipley (2006) found support for the persistence, position updating, and relatability notions of STR. The notion of persistence was supported because observers performed as if they had seen dynamically occluded objects for longer than their physical exposure durations. The notion of position updating was supported because observers were highly accurate at discriminating between two shape configurations that differed only in the horizontal alignment of the pieces. Because the dynamically occluded objects traveled horizontally and two partner edges on either side of an occluded region were not seen simultaneously, observers' accurate performance demonstrated that they had information about the locations of both edges despite the fact that, at all times, at least one was occluded. Finally, the notion that contour and surface interpolation processes operated in these displays was supported by a strong advantage in discrimination performance under conditions predicted to support object formation. Specifically, configurations that conformed to the geometric constraints of STR produced markedly better sensitivity than those that did not. The dependence of this effect on object formation was also shown in another condition, in which the removal of a mere 6% of pixels at the points of occlusion (rounding contour junctions) produced reliably poorer performance, despite the fact that the global configuration of the pieces was preserved. This last result was predicted from prior findings that rounding of contour junctions weakens contour completion processes (Albert, 2001; Shipley and Kellman, 1990; Rubin, 2001). When the contour interpolation process was compromised, the three projected fragments of the objects were less likely to be perceived as a single visual unit, and discrimination performance suffered.

Palmer, Kellman, and Shipley (2006) proposed the notion of a *dynamic visual icon*—a representation in which the persistence and position updating (and perhaps interpolation) functions of STR are carried out. The idea extends the notion of a visual icon representation, first discovered by Sperling (1960) and labeled by Neisser (1967), which accepts information over time and allows the perceiver to integrate visual information that is no longer physically visible. Perception is not an instantaneous process but rather is extended over time, and the visual icon is a representation that faithfully maintains visual information in a spatially accurate format for 100 ms or more after it disappears. The proposal of a dynamic visual icon adds the idea that represented information may be positionally updated based on previously acquired velocity information. It is not clear whether this feature is a previously unexplored aspect of known iconic visual representations or whether it implicates a special representation. What is clear is that visual information is not only accumulated over time and space, but that the underlying representation is geared toward the processing of ongoing events.

A yet unexplored aspect of the dynamic visual icon is whether it incorporates position change information in all three spatial dimensions. If so, this sort of representation might handle computations in a truly 4D spatiotemporal object formation process. Most studies to date, as well as the theory of STR proposed by Palmer, Kellman, and Shipley (2006), have focused on 2D contour completion processes and motion information. A more comprehensive idea of 3D interpolation that incorporates position change and integration over time has yet to be studied experimentally. Future work will address this issue and attempt to unify the 3D object formation work of Kellman, Garrigan, and Shipley (2005) with the STR theory and findings of Palmer, Kellman, and Shipley (2006).

## FROM SUBSYMBOLIC TO SYMBOLIC REPRESENTATIONS

One way to further our understanding of contour interpolation processes is to build models of contour interpolation mechanisms and compare their performance to human perception. One such model takes grayscale images as input, and using simulated simple and complex cells, detects contour junctions (Heitger et al., 1992) and interpolates between them using geometric constraints much like contour relatability (Heitger et al., 1998).

The output of this model is an image of activations at pixel locations along interpolation paths. These activations appear at locations in images where people report perceiving illusory contours.

A more recent model (Kalar et al., 2010), generalizes the framework proposed in the model of Heitger et al. (1998) to interpolate both illusory and occluded contours. This model, which is a neural implementation of contour relatability and the identity hypothesis (that illusory and occluded objects share a common underlying interpolation mechanism), generates images of illusory and occluded contours consistent with human perception in a wide variety of contexts (e.g., Figure 10.12).

There is, however, an important shortcoming of models of this type. The inputs to these models are *images* where pixel values represent luminance. The outputs of the models are *images* of illusory and occluded (and real) contours. That is, these models take images that represent luminance at each pixel location and return images that represent illusory and occluded interpolation *activity* at each pixel location. There is nothing in the outputs to indicate that different pixels are connected to each other, that they form part of a contour, and so on. Nothing describes the areas that form a complete object, much less provides a description of its shape. In this sense, such models can be easily misinterpreted, as the observer viewing the outputs provides all of these additional descriptions.

(a)                                      (b)

FIGURE 10.12 Displays and outputs from a filtering model that uses a unified operator to handle illusory and occluded interpolations. The model (Kalar et al., 2010) draws heavily on Heitger et al. (1998) but replaces their "ortho" and "para" grouping processes by a single operator sensitive to either L or T junction inputs. (a) Kanizsa-style transparency display on the left produces output on the right. The illusory contours would not be interpolated by a model sensitive to L junctions only (e.g., Heitger et al., 1998). Except for triangle vertices, all junctions in this display are anomalous T junctions. (b) Occlusion display on the left produces output on the right. This output differs from the Heitger et al. model, which is intended to interpolate only illusory contours.

To account for human perception, the important filtering information provided by these models must feed into mechanisms that produce higher-level, symbolic descriptions. Beyond representations of contours (both physically defined and interpolated) as sets of unbound pixel values must be a more holistic description, with properties like shape, extent, and their geometric relationships to other contours in the scene.

This is a very general point about research into object and surface perception. Vision models using relatively local filters describe important basic aspects of processing. At a higher level, some object recognition models assume tokens such as contour shapes, aspect ratios, or volumetric primitives as descriptions. The difficult problem is in the middle: How do we get from local filter responses to higher-level symbolic descriptions? This question is an especially high priority for understanding 3D shape and surface perception. Higher-level shape descriptions are needed to account for our use of shape in object recognition and our perceptions of similarity. As the Gestalt psychologists observed almost a century ago, two shapes may be seen as similar despite being composed of very different elements. Shape cannot be the sum of local filter activations. Understanding processes that bind local responses into unitary objects and achieve more abstract descriptions of these objects are crucial challenges for future research.

## 3D PERCEPTION IN THE BRAIN

Research in 3D and spatiotemporal perception also raises important issues for understanding the cortical processes of vision. A great deal of research and modeling has focused on early cortical areas, V1 and V2, as likely sites of interpolation processes (Mendola et al., 1999; Sugita, 1999; Bakin, Nakayama, and Gilbert, 2000; Seghier et al., 2000) for both illusory and occluded contours. On the basis of their results on 3D interpolation, Kellman, Garrigan, and Shipley (2005) suggested that interpolation processes involve all three spatial dimensions and are unlikely to be accomplished in these early areas. There are several reasons for this suggestion. First, orientation-sensitive units in V1 and V2 encode 2D orientation characteristics, which are not sufficient to account for 3D interpolation. Second, one could hypothesize that 2D orientations combined with outputs of disparity-sensitive cells might somehow provide a basis for 3D interpolation. Evidence indicates, however, that the type of disparity information available in these early areas is insufficient: whereas relative disparities are needed for depth computations, V1 neurons with disparity sensitivity appear to be sensitive to absolute disparities, which vary with fixation (Cumming and Parker, 1999). Third, even relative disparities do not directly produce perception of depth intervals in the world. There are two problems. One is that obtaining a depth interval from disparity involves a constancy problem; a given depth interval produces different disparity differences depending on viewing distance (Wallach and Zuckerman, 1963). To obtain a depth interval, disparity information must be combined with egocentric distance information, obtained from some other source, to at least one point. Figure 10.13 illustrates this problem, along with a second one. The experimental data of Kellman et al. (2005) suggest that edge segments at particular slants provide the inputs to 3D interpolation processes. Slant, however, depends not only on the depth interval between points, but also on their separation. Moreover, 3D slant may be specified from a variety of sources. The likely substrate of 3D interpolation is some cortical area in which actual slant information, computed from a variety of contributing cues, is available. These requirements go well beyond computations that are suspected to occur in V1 or V2.

Where might such computations take place? Although no definite neural locus has been identified, research using a single-cell recording in the

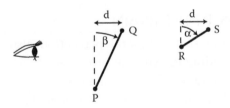

FIGURE 10.13 Relations between disparity, edge length, and slant. *Depth and disparity*: A given depth interval *d* in the world gives rise to decreasing disparity as viewing distance increases. Interval d given by points R and S produces smaller disparity differences than P and Q, if R and S are farther away. *Slant*: Obtaining slant from disparity depends not only on recovering the depth interval but also depends on the vertical separation of the points defining that depth interval ($\alpha > \beta$).

caudal intraparietal sulcus (cIPS) indicates the presence of units tuned to 3D slants (Sakata et al., 1997). It is notable that these units appear to respond to a particular 3D orientation regardless of whether that orientation is specified by stereoscopic information or texture information. These findings indicate where the kinds of input units required for 3D relatability—namely, units signaling 3D orientation and position—may exist in the nervous system.

In addition to the location of 3D processing, much remains to be learned about the nature of its mechanisms. Are there areas of cortex in which units sensitive to 3D positions and orientations of contour or surface fragments interact in a network that represents 3D relations? At present, we do not know of such a network, but the psychophysical results suggest that it is worth looking for.

All of these same sorts of questions apply as well to spatiotemporal object formation. We have impressive capabilities to construct coherent objects and scenes from fragments accumulated across gaps in space and time. Where in the cortex are these capabilities realized? And what mechanisms carry out the storage of previously visible fragments and their positional updating, based on velocity information, after they have gone out of sight? A striking possibility is that the same cortical areas are involved as those in 3D interpolation. At least, such an outcome would be consistent with a grand unification of processes that create objects from fragments. Although individual experiments have usually addressed 2D, 3D, and spatiotemporal interpolation separately, they may be part of a more comprehensive 4D process. Understanding both the computations involved and their neural substrates are fundamental and exciting issues for future research.

# CHAPTER 11

# The Perceptual Representation of 3D Shape

James T. Todd

## CONTENTS

## INTRODUCTION

Human observers have a remarkable ability to determine the three-dimensional (3D) structures of objects in the environment based on patterns of light that project onto the retina. There are several different aspects of optical stimulation that are known to provide useful information about the 3D layout of the environment, including shading, texture, motion, and binocular disparity. Although numerous computational models have been developed for estimating 3D structure from these different sources of optical information, effective algorithms for analyzing natural images have proven to be surprisingly elusive. One possible reason for this, I suspect, is that there has been relatively little research to identify the specific aspects of an object's structure that form the primitive components of an observer's perceptual knowledge. After all, in order to compute shape, it is first necessary to define what "shape" is.

An illuminating discussion about the concept of shape was first published almost 50 years ago by the theoretical geographer William Bunge

(1962). Bunge argued that an adequate measure of shape must satisfy four criteria: it should be objective; it should not include less than shape, such as a set of position coordinates; it should not include more than shape, such as fitting it with a Fourier or Taylor series; and it should not do violence to our intuitive notions of what constitutes shape. This last criterion is particularly important for understanding human perception, and it highlights the deficiencies of most 3D representations that have been employed in the literature. For example, all observers would agree that a big sphere and a small sphere both have the same shape—a sphere—but these objects differ in almost all of the standard measures with which 3D structures are typically represented (see Koenderink, 1990). In this chapter, I will consider two general types of data structures—maps and graphs—that are used for the representation of 3D shape by virtually all existing computational models, and I will also consider their relative strengths and weaknesses as models of human perception.

## LOCAL PROPERTY MAPS

Consider the image of a smoothly curved surface that is presented in Figure 11.1. Clearly there is sufficient information from the pattern of shading to produce a compelling impression of 3D shape, but what precisely do we know about the depicted object that defines the basis of our perceptual representations? Almost all existing theoretical models for computing the 3D structures of arbitrary surfaces from shading, texture, motion, or binocular disparity are designed to generate a particular form of data structure that we will refer to generically as a *local property map*. The basic idea is quite simple and powerful. A visual scene is broken up into a matrix of small local neighborhoods, each of which is characterized by a number (or a set of numbers) to represent some particular local aspect of 3D structure.

The most common variety of this type of data structure is a depth map, in which each local region of a surface is defined by its distance from the point of observation. Although depth maps are used frequently for the representation of 3D shape, they have several undesirable characteristics. Most importantly, they are extremely unstable to variations in position, orientation, or scale. Because these transformations have a negligible impact on the perception of 3D shape, we would expect that our underlying perceptual representations should be invariant to these changes as well, but depth maps do not satisfy that criterion.

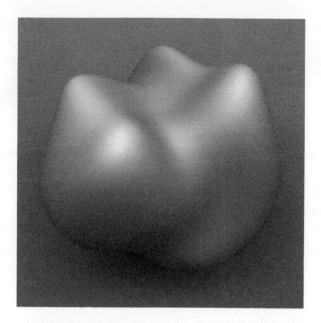

FIGURE 11.1  A shaded image of a smoothly curved surface. What do observers know about this object that defines the basis of their perceptual representations?

The stability of local property maps can be improved somewhat by using higher levels of differential structure to characterize each local region. A depth map, $Z = f(X,Y)$ represents the 0th order structure of a surface, but it is also possible to define higher-order properties by taking spatial derivatives of this structure in orthogonal directions. The first spatial derivatives of a depth map $(f_X, f_Y)$ define a pattern of surface depth gradients (i.e., an orientation map). Similarly, its second spatial derivatives $(f_{XX}, f_{XY}, f_{YY})$ define the pattern of curvature on a surface. Note that these higher levels of differential structure can be parameterized in a variety of ways. For example, many computational models represent local orientations using the surface depth gradients in the horizontal and vertical directions $(f_X, f_Y)$, which is sometimes referred to as gradient space. An alternative possibility is to represent surface orientation in terms of *slant* ($\sigma$) and *tilt* ($\tau$), which are equivalent to the latitude and longitude, respectively, on a unit sphere.

There are also different coordinate systems for representing second-order differential structure. One common procedure is to specify the maximum and minimum curvatures $(k_1, k_2)$ at each local region, which are always orthogonal to one another. Another possibility is to parameterize the structure in terms of *mean curvature* ($H$) and *Gaussian curvature* ($K$), where $H = (k_1 + k_2)/2$ and $K = k_1 k_2$ (see Tyler, 2010). Perhaps the most perceptually plausible representation of local curvature was proposed by Koenderink (1990), in which the local surface structure is parameterized in terms of *curvedness* ($C$) and *shape index* ($S$), where $C = \sqrt{k^2 + k_2^2}$ and $S = \arctan 2(k_1, k_2)$. An important property of this latter representation is that curvedness varies with object size, whereas the shape index component does not. Thus, a point on a large sphere would have a different curvedness than would a point on a small sphere, but they would both have the same shape index.

There have been many different studies reported in the literature that have measured the abilities of human observers to estimate the local properties of smooth surfaces (see Figure 11.2). One approach that has been employed successfully in numerous experiments is to present randomly generated surfaces, such as the one shown in the left panel of Figure 11.2, with two small probe points to designate the target regions whose local properties must be compared. This technique has been used to investigate observers' perceptions of both relative depth (Koenderink, van Doorn, and Kappers, 1996; Todd and Reichel, 1989; Norman and Todd, 1996, 1998) and relative orientation (Todd and Norman, 1995; Norman and Todd, 1996) under a wide variety of conditions. One surprising result that has been obtained repeatedly in these studies is that the perceived relative depth of two local regions can be dramatically influenced by the intervening surface structure along which they are connected. It is as if the shape of a visible surface can produce local distortions in the structure of perceived space. One consequence of these distortions is that observers' sensitivity to relative depth intervals can deteriorate rapidly as their spatial separation along a surface is increased. Interestingly, these effects of separation are greatly diminished if the probe points are presented in empty space, or for judgments of relative orientation intervals (see Todd and Norman, 1995; Norman and Todd, 1996, 1998).

Another technique for measuring observers' perceptions of local surface structure involves adjusting the 3D orientation of a circular disk, called a gauge figure, until it appears to rest in the tangent plane at some designated surface location (Koenderink, van Doorn, and Kappers, 1996; Todd et al., 1997). There are several variations of this technique that have been adapted for different types of optical information. For stereoscopic displays, the gauge figure is presented monocularly so that the adjustment

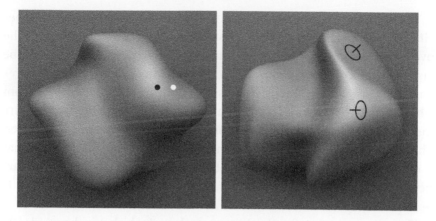

FIGURE 11.2   Alternative methods for the psychophysical measurement of perceived three-dimensional shape. The left panel depicts a possible stimulus for a relative depth probe task. On each trial, observers must indicate by pressing an appropriate response key whether the black dot or the white dot appears closer in depth. The right panel shows a common procedure for making judgments about local surface orientation. On each trial, observers are required to adjust the orientation of a circular disk until it appears to fit within the tangent plane of the depicted surface. Note that the probe on the upper right of the object appears to satisfy this criterion, but that the one on the lower left does not. By obtaining multiple judgments at many different locations on the same object, it is possible with both of these procedures to compute a specific surface that is maximally consistent in a least-squares sense with the overall pattern of an observer's judgments.

cannot be achieved by matching the disparities of nearby texture elements. Although the specific depth of the gauge figure is mathematically ambiguous in that case, most observers report that it appears firmly attached to the surface, and that they have a high degree of confidence in their adjustments. A similar technique can also be used with moving displays. However, to prevent matches based on the relative motions of nearby texture elements, the task must be modified somewhat so that observers adjust the shape of an ellipse in the tangent plane until it appears to be a circle (Norman, Todd, and Phillips, 1995). This latter variation has also been used for the direct viewing of real objects by adjusting the shape of an ellipse that is projected on a surface with a laser beam (Koenderink, van Doorn, and Kappers, 1995).

Jan Koenderink developed a toolbox of procedures for computing a specific surface that is maximally consistent in a least-squares sense with the overall pattern of an observer's judgments for a variety of different probe tasks. The reconstructed surfaces obtained through these procedures almost always appear qualitatively similar to the stimulus objects from which they were generated, although it is generally the case that they are metrically distorted. These distortions are highly constrained, however. They typically involve a scaling or shearing transformation in depth that is consistent with the family of possible of 3D interpretations for the particular source of optical information with which the depicted surface is perceptually specified (see Koenderink et al., 2001).

To summarize briefly, the literature described above shows clearly that human observers can make judgments about the geometric structure (e.g., depth, orientation, or curvature) of small, local, surface regions. Depending on the particular probe task employed, these judgments can sometimes be reasonably reliable for individual observers (see Todd et al., 1997). However, observers' judgments often exhibit large systematic distortions relative to the ground truth, and there can also be large differences among the patterns of judgments obtained from different observers. It is important to point out that these perceptual distortions are almost always limited to a scaling or shearing transformation in depth. Thus, although observers seem to have minimal knowledge about the metric structures of smoothly curved surfaces, they are remarkably accurate in judging the more qualitative properties of 3D shape that are invariant over those particular affine transformations that define the space of perceptual ambiguity. (See Koenderink et al., 2001, for a more detailed discussion.)

## FEATURE GRAPHS

It is important to note that all of the local properties described thus far are metrical in nature. That is to say, it is theoretically possible for each visible surface point to be represented with appropriate numerical values for its depth, orientation, or curvature, and these values can be scaled in an appropriate space so that the distances between different points can be easily computed. However, there are other relevant aspects of local surface structure that are fundamentally nonmetrical and are best characterized in terms of ordinal or nominal relationships. Consider, for example, the drawing of an object presented in Figure 11.3. There is an extensive literature in both human and machine vision that has been devoted to the

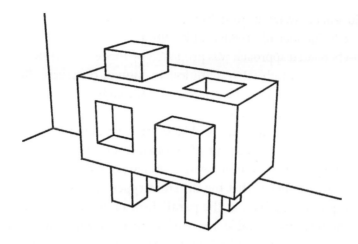

FIGURE 11.3   Line drawings of objects can provide convincing pictorial representations even though they effectively strip away almost all of the variations in color and shading that are ordinarily available in natural scenes.

analysis of this type of polyhedral object (e.g., see Waltz, 1975; Winston, 1975; Biederman, 1987). A common form of representation employed in these analyses is focused exclusively on localized features of the depicted structure and their ordinal and topological relations (e.g., above or below, connected or unconnected). These features are typically of two types: point singularities called *vertices*, and line singularities called *edges*. The pattern of connectivity among these features is referred to mathematically as a graph.

The first rigorous attempts to represent the structure of objects using feature graphs were developed in the 1970s by researchers in machine vision (e.g., Clowes, 1971; Huffman, 1971, 1977; Mackworth, 1973; Waltz, 1975). An important inspiration for this early work was the phenomenological observation from pictorial art that it is possible to convey the full 3D shape of an object by drawing just a few lines to depict its edges and occlusion contours, thus suggesting that a structural description of these basic features may provide a potentially powerful representation of an object's 3D structure. To demonstrate the feasibility of this approach, researchers were able to exhaustively catalog the different types of vertices that can arise in line drawings of simple plane-faced polyhedra, and then used that to label which lines in a drawing correspond to convex, concave, or occluding edges. Similar procedures were later developed to deal with the occlusion contours of smoothly curved surfaces (Koenderink and van Dorn, 1982; Koenderink, 1984; Malik, 1987).

A closely related approach was proposed by Biederman (1987) for modeling the process of object recognition in human observers. Biederman argued that most namable objects can be decomposed into a limited set of elementary parts called *geons*, and that the arrangement of these geons can adequately distinguish most objects from one another, in much the same way that the arrangement of phonemes or letters can adequately distinguish spoken or written words. A particularly important aspect of Biederman's theory is that the different types of geons are defined by structural properties, such as the parallelism or cotermination of edges, which remain relatively stable over variations in viewing direction. Thus, the theory can account for a limited degree of viewpoint invariance in the ability to recognize objects across different vantage points (see also Tarr and Kriegman, 2001).

In principle, the edges in a graph representation can have other attributes in addition to their pattern of connectivity. For example, they are often labeled to distinguish different types of edges, such as smooth occlusion contours, or concave or convex corners. They may also have metrical attributes, such as length, orientation, or curvature. Lin Chen (1983, 2005) proposed an interesting theoretical hypothesis that the relative perceptual salience of these different possible edge attributes is systematically related to their structural stability under change, in a manner that is similar to the Klein hierarchy of geometries. According to this hypothesis, observers should be most sensitive to those aspects of an object's structure that remain invariant over the largest number of possible transformations, such that changes in topological structure (e.g., the pattern of connectivity) should be more salient than changes in projective structure (e.g., straight versus curved), which, in turn, should be more salient than changes in affine structure (e.g., parallel versus nonparallel). The most unstable properties are metric changes in length, orientation, or curvature. Thus, it follows from Chen's hypothesis that observers' judgments of these attributes should produce the lowest levels of accuracy or reliability.

There is some empirical evidence to support that prediction. For example, Todd, Chen, and Norman (1998) measured accuracy and reaction times for a match to sample task involving stereoscopic wire frame figures in which the foil could differ from the sample in terms of topological, affine, or metric structure. The relative performance in these three conditions was consistent with Chen's predictions. Observers were most accurate

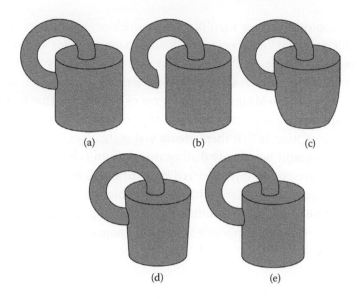

FIGURE 11.4 A base object on the upper left (a), and several possible distortions of it that alter different aspects of the geometric structure; (b) a topological change; (c) a projective change; (d) an affine change; and (e) a metric change. All of these changes have been equated with respect to the number of pixels that have been altered relative to the base object.

and had the fastest response times when the foil was topologically distinct from the sample, and they were slowest and least accurate when the only differences between the foil and standard involved metrical differences of length or orientation. A similar result was also obtained by Christensen and Todd (2006) using a same–different matching task involving objects composed of pairs of geons like those shown in Figure 11.4. Thresholds for different types of geometric distortions were measured as a function of the number of changed pixels in an image that was needed to detect that two objects had different shapes. The ordering of these thresholds was again consistent with what would be expected based on the Klein hierarchy of geometries (see also Biederman and Bar, 1999). Still other evidence to confirm this pattern of results was obtained in shape discrimination tasks of wire-frame figures that are optically specified by their patterns of motion (Todd and Bressan, 1990).

The relative stability of image structure over changing viewing conditions is a likely reason why shape discrimination by human observers is primarily based on the projected pattern of an object's edges and occlusion contours, rather than pixel intensities or the outputs of simple cells in the primary visual cortex. The advantage of edges and occlusion contours for the representation of object shape is that they are invariant over changes in surface texture or the pattern of illumination (e.g., see Figure 11.3), thus providing a greater degree of constancy. That being said, however, an important limitation of this type of representation as a model of human perception is that there are no viable theories at present of how these features could be reliably distinguished from other types of image contours that are commonly observed in natural vision, such as shadows, reflectance edges, or specular highlights. The solution to that problem is a necessary prerequisite for successfully implementing an edge-based representation that would be applicable to natural images in addition to simple line drawings.

# References

Adams WJ, Mamassian P (2004) Bayesian combination of ambiguous shape cues. *J Vision* 4:921–929. doi:10.1167/4.10.7.

Adelson EH, Freeman WT (1991) The design and use of steerable filters. *IEEE Trans Patt Anal Mach Intell (PAMI)* 13:891–906.

Adiv G (1985) Determining 3D motion and structure from optical flow generated by several moving objects. *IEEE Trans Patt Anal Mach Intell (PAMI)* 7:384–401.

Adiv G (1989) Inherent ambiguities in recovering 3D motion and structure from a noisy flow field. *IEEE Trans Patt Anal Mach Intell (PAMI)* 11:477–489.

Albert MK (2001) Surface perception and the generic view principle. *Trends Cogn Sci* 5:197–203.

Albert MK, Hoffman DD (2000) The generic-viewpoint assumption and illusory contours. *Perception* 29:303–312.

Albert MK, Tse PU (2000) The role of surface attraction in perceiving volumetric shape. *Perception* 29:409–420.

Aloimonos J, Weiss I, Bandyopadhyay A (1988) Active vision. *Int J Comput Vision* 1:333–356.

Alpert S, Galun M, Basri R, Brandt A (2007) Image segmentation by probabilistic bottom-up aggregation and cue integration. In *Proc IEEE Comput Vision Patt Recogn (CVPR)* 23:1–8.

Antoine JP (2000) Coherent states, wavelets and beyond. In *Contemporary Problems in Mathematical Physics* (Cotonou, 1999). River Edge, NJ: World Sci. Publishing, 3–46.

Arbelaez P, Cohen L (2008) Constrained image segmentation from hierarchical boundaries. In *Proc IEEE Comput Vision Patt Recogn (CVPR)* 24:454–467.

Ashley M (1898) Concerning the significance of light in visual estimates of depth. *Psych Rev* 5:595–615.

Aubin T (1976) Equations différéntielles non linéaires et problème de Yamabe concernant la courbure scalaire. *J Math Pures Appl* 55:269–296.

August J, Zucker SW (2003) Sketches with curvature: The curve indicator random field and markov processes. *IEEE Trans Patt Anal Mach Intell* (PAMI) 25:387–400.

Ayer S, Schroeter P, Bigfin J (1994) Segmentation of moving objects by robust motion parameter estimation over multiple frames. In *Proc ECCV*, Berlin: Springer-Verlag, 316–327.

Bagon S, Boiman O, Irani M (2008) What is a good image segment? A unified approach to segment extraction. In Forsyth D, Torr P, Zisserman A (eds) *Computer Vision—ECCV 2008, LNCS*, Springer-Verlag 5305:30–44.

Bajcsy R (1988) Active perception. *Proc. IEEE, Computer Vision*, 76:966–1005.

Bakin JS, Nakayama K, Gilbert C (2000) Visual responses in monkey areas V1 and V2 to three-dimensional surface configurations. *J Neurosci* 21:8188–8198.

Banton T, Levi DM (1992) The perceived strength of illusory contours. *Percept Psychophys* 52:676–684.

Barbieri D, Citti G, Sanguinetti G, Sarti A (2010) An uncertainty principle underlying the pinwheels structure in the primary visual cortex, *arXiv* 1007.1395

Barrett WA, Mortensen EN (1997) Interactive live-wire boundary extraction. *Medical Image Analysis*, 1:331–341.

Barrow HG, Tenenbaum JM (1981) Interpreting line drawings as three-dimensional surfaces. *Artific Intell* 17:75–116.

Battaglia PW, Jacobs RA, Aslin RN (2003) Bayesian integration of visual and auditory signals for spatial localization. *J Opt Soc Am A* 20:1391–1397.

Baylis GC, Driver J (1993) Visual attention and objects: Evidence for hierarchical coding of location. *J Exp Psych: Human Percept Perf* 19:451–470.

Ben Shahar O, Zucker SW (2003) Geometrical computations explain projection patterns of long-range horizontal connections in visual cortex. *Neural Comput* 16:445–476.

Biederman I (1987) Recognition-by-components: A theory of human image understanding. *Psych Rev* 94:115–147.

Biederman I, Bar M (1999) One-shot viewpoint invariance in matching novel objects. *Vision Res* 39:2885–2889.

Blake A, Bülthoff HH, Sheinberg D (1993) Shape from texture: Ideal observer and human psychophysics. *Vision Res* 33:1723–1737.

Blake A, Rother C, Brown M, Perez P, Torr P (2004) Interactive image segmentation using an adaptive GMMRF model. In *Eur Conf Comput Vision (ECCV)*, 428–441.

Bobenko AI, Springborn BA (2004) Variational principles for circle patterns and Koebe's theorem. *Trans Amer Math Soc* 356:659–689.

Bober M, Kittler J (1994) Robust motion analysis. In *Proc IEEE Comput Vision Patt Recogn (CVPR)* 10:947–952.

Bosking W, Zhang Y, Schoenfield B, Fitzpatrick D (1997) Orientation selectivity and the arrangement of horizontal connections in tree shrew striate cortex. *J Neurosci* 17:2112–2127.

Bowers PL, Hurdal MK (2003) Planar conformal mapping of piecewise flat surfaces. In *Visualization and Mathematics III*, Hege HC, Polthier K (eds), Springer-Verlag: Berlin, 3–34.

Boykov YY, Kolmogorov V (2004) An experimental comparison of min-cut/max-flow algorithms for energy minimization in vision. *IEEE Trans Patt Anal Mach Intell (PAMI)* 26:359–374.

Boykov YY, Jolly JP (2001) Interactive graph cuts for optimal boundary and region segmentation of objects in n-D images. In *IEEE Int Conf Comput Vision* 1:105–112.

Boykov Y, Veksler O, Zabih R (2001) Fast approximate energy minimization via graph cuts. *IEEE Trans Patt Anal Mach Intell (PAMI)* 23:1222–1239.

Braddick O (1974) A short-range process in apparent motion. *Vision Res* 14:519–527.

Bradley DC (2001) Early visual cortex: Smarter than you think. *Curr Biol* 11:95–98.

Bredfeldt CE, Cumming BG (2006) A simple account of cyclopean edge responses in macaque V2. *J Neurosci* 26:7581–7596.

Bressloff PC, Cowan JD, Golubitsky M, Thomas PJ, Wiener M (2001) Geometric visual hallucinations: Euclidean symmetry and the functional architecture of striate cortex. *Phil Trans Roy Soc B* 40:299–330.

Buckley D, Frisby JP, Blake A (1996) Does the human visual system implement an ideal observer theory of slant from texture? *Vision Res* 36:1163–1176.

Bunge W (1962) *Theoretical Geography.* C. W. K. Gleerup: Lund, Sweden.

Burt P, Wixson L, Salgian, G (1995) Electronically direted "focal" stereo. In: *International Conference on Computer Vision* (ICCV) 95:94–101

Burt PJ, Bergen JR, Hingorani R, Kolczynski R, Lee WA, Leung A, Lubin J, Shvayster H (1989) Object tracking with a moving camera. In *IEEE Proc Workshop Visual Motion*, 2–12.

Carandini M, Ringach DL (1997) Predictions of a recurrent model of orientation selectivity. *Vision Res* 37:3061–3071.

Carman G J, Welch L (1992). Three dimensional illusory contours and surfaces. *Nature* 360:585–587.

Carmichael HJ (2002) *Statistical Methods in Quantum Optics 1: Master Equations and Fokker Planck Equations,* Springer: Berlin.

Carner C, Jin M, Gu X, Qin H (2005) Topology driven surface mappings with robust feature alignment. *IEEE Trans Visualiz Comput Graphics (VCG)* 16:543–550.

Carr H (1935) *An Introduction to Space Perception.* Longmans Green: London.

Cerf M, Harel J, Einhauser W, Koch C (2008) Predicting human gaze using low-level saliency combined with face detection. In *Adv Neural Inf Proc Syst (NIPS)* 20:241–248.

Chan MW, Stevenson AK, Li Y, Pizlo Z (2006) Binocular shape constancy from novel views: The role of a priori constraints. *Percept Psychophys* 68:1124–1139.

Chellappa R, Jain A (eds) (1993) *Markov Random Fields: Theory and Application.* Academic Press: Boston.

Chen L (1983) Topological structure in visual perception. *Science* 218:699–700.

Chen L (2005) The topological approach to perceptual organization. *Visual Cogn* 12:553–637.

Chow B (1991) The Ricci flow on the 2-sphere. *Journal of Differential Geometry* 33:325–334.

Chow B, Luo F (2003) Combinatorial Ricci flows on surfaces. *J Diff Geom* 63:97–129.

Christensen HI, Eklundh JO (2000) Active vision from multiple cues. In (*BMVC*) *'00: Proc 1st IEEE Int Workshop Biol Motiv Comput Vision (BMVC)*, London UK: Springer-Verlag 209–216.

Christensen J, Todd JT (2006) What image measures are best correlated with the discriminability of 3D objects? *J Vision* 6:321 doi: 10.1167/6.6.321.

Citti G, Sarti A (2006) A cortical based model of perceptual completion in the roto-translation space. *J Math Imaging Vision* 24:307–326.

Clark JJ, Yuille AL (1990) *Data Fusion for Sensory Information Processing Systems.* Kluwer Academic: Dordrecht.

Clowes MB (1971) On seeing things. *Artif Intell* 2:79–116.

Cochran WG (1937) Problems arising in the analysis of a series of similar experiments. *J Roy Stat Soc, Suppl* 4:102–118.

Collins C, Stephenson K (2003) A circle packing algorithm. *Comput Geometry: Theory Appl* 25:233–256.

Costeira J, Kanade T (1995) A multi-body factorization method for motion analysis. In *Int Conf Comput Vision* 70:41–54.

Coughlan JM (2011) Mechanisms for propagating surface information in 3-D reconstruction. In Tyler CW (ed) (2010) *Computer Vision: From Surfaces to 3D Objects*, Boca Raton, FL: Taylor & Francis, Ch 3.

Coules J (1955) Effect of photometric brightness on judgments of distance. *J Exp Psychol* 50:19–25.

Craft E, Schuetze H, Niebur E, von der Heydt R (2007) A neural model of figure–ground organization. *J Neurophysiol* 97: 4310–4326.

Cryer JE, Tsai PS, Shah M (1995) Integration of shape from shading and stereo. *Patt Recogn* 28:1033–1043.

Cumming BG, Johnston EB, Parker AJ (1993) Effects of different texture cues on curved surfaces viewed stereoscopically. *Vision Res* 33:827–838.

Cumming BG, Parker AJ (1999) Binocular neurons in V1 of awake monkeys are selective for absolute, not relative, disparity. *J Neurosci* 19:5602–5618.

Cutting JE, Vishton PM (1995) Perceiving layout and knowing distances: The integration, relative potency, and contextual use of different information about depth. In Epstein W, Rogers SJ (eds) *Perception of Space and Motion, Handbook of Perception and Cognition*. Academic Press: San Diego, 69–117.

Daugman JG (1985) Uncertainty relation for resolution in space, spatial frequency and orientation optimized by two dimensional visual cortical filters. *J Opt Soc Amer A* 2:1160–1169.

De Angelis GC, Ohzawa I, Freeman RD (1995) Receptive-field dynamics in the central visual pathways. *Trends Neurosci* 18:451–458.

Doya K, Ishii S, Pouget A, Rao R (eds) (2006) *The Bayesian Brain: Probabilistic Approaches to Neural Coding*. MIT Press: Cambridge.

Duits R, Franken E (2007) Left-invariant stochastic evolution equations on se(2) and its applications to contour enhancement and contour completion via invertible orientation scores. *ArXiv* 0711.0951v4.

Durand JB, Nelissen K, Joly O, Wardak C, Todd JT, Norman JF, Janssen P, Vanduffel W, Orban GA (2007) Anterior regions of monkey parietal cortex process visual 3D shape. *Neuron* 55:493–505.

Durand JB, Peeters R, Norman JF, Todd JT, Orban GA (2009) Parietal regions processing visual 3D shape extracted from disparity. *NeuroImage* 46:1114–1126.

Ernst MO, Banks MS (2002) Humans integrate visual and haptic information in a statistically optimal fashion. *Nature* 415:429–433.

Ernst MO, Banks MS, Bülthoff HH (2000) Touch can change visual slant perception. *Nature Neurosci* 3:69–73.

Fantoni C, Gerbino W (2003) Contour interpolation by vector-field combination. *J Vision* 3:281–303.

Fantoni C, Hilger JD, Gerbino W, Kellman PJ (2008) Surface interpolation and 3D relatability. *J Vision* 8:29, 1–19, http://journalofvision. org/8/7/29/, doi: 10. 1167/8. 7. 29.

Farne M (1977) Brightness as an indicator to distance: Relative brightness per se or contrast with the background? *Perception* 6:287–293.

Fechner G (1860/1966) *Elements of Psychophysics*. Holt, Rinehart & Winston: New York.

Felzenszwalb P, Huttenlocher D (2006) Efficient belief propagation for early vision *Int J Comput Vision* 70:41–54.

Field DJ (1994) What is the goal of sensory coding? *Neural Comput* 6:559–601.

Field DJ, Hayes A, Hess RF (1993) Contour integration by the human visual system: Evidence for a local association field. *Vision Res* 33:173–193.

Folland GB (1989) *Harmonic Analysis in Phase Space*. Princeton, NJ: Princeton Univ Press.

Fowlkes CC, David RM, Malik J (2007) Local figure/ground cues are valid for natural images. *J Vision*, 7: 1–9 doi:10.1167/7.8.2.

Franken E, Duits R, ter Haar Romeny BM (2007) Nonlinear diffusion on the 2nd Euclidean motion group. In Sgallari F, Murli A, Paragios N (eds) SSVM, *Lect Notes Comput Sci*, Springer, Berlin, 4485:461–472.

Freeman WT, Pasztor E, Carmichael OT (2000) Learning low-level vision. *Int J Comput Vision* 40:25–47.

Freeman WT, Torralba A (2003) Shape recipes: Scene representations that refer to the image. In *Adv Neural Inf Proc Syst (NIPS) 15*, MIT Press: Cambridge, MA.

Frey BJ (1998) *Graphical Models for Machine Learning and Digital Communication*. MIT Press: Cambridge, MA.

Frey BJ, Dueck, D (2007) Clustering by passing messages between data points. *Science*, January.

Frey BJ, Koetter R, Petrovic N (2002) Very loopy belief propagation for unwrapping phase images. In *Advances in Neural Information Processing Systems 14*. Dietterich TG, Becker S, Ghahramani Z (eds), MIT Press: Cambridge, MA, 737–744.

Friedman, JH, Stuetzle, W, Schroeder, A (1984). Projection pursuit density estimation. *Journal of the American Statistical Association* 79:599–608.

Friedman HS, Zhou H, von der Heydt R (1999) Color filling-in under steady fixation: Behavioral demonstration in monkeys and humans. *Perception* 28: 1383–1395.

Friedman HS, Zhou H, von der Heydt R (2003) The coding of uniform color figures in monkey visual cortex. *J Physiol (Lond)* 548:593–613.

Gabor D (1946) Theory of communication, *J IEEE* 93:429–459.

Geisler WS (2003) Ideal observer analysis. In *The Visual Neurosciences*, Chalupa L, Werner J (eds), MIT Press: Boston, 825–837.

Geisler WS, Perry JS, Super BJ, Gallogly DP (2001) Edge co-occurrence in natural images predicts contour grouping performance. *Vision Res* 41:711–724.

Georgieva S, Peeters R, Kolster H, Todd JT, Orban GA (2009) The processing of three-dimensional shape from disparity in the human brain. *J Neurosci* 29:727–742.

Gershenfeld N (1998) *The Nature of Mathematical Modeling*. Cambridge University Press: Cambridge.

Gibson JJ (1950) *The Perception of the Visual World*. Houghton Mifflin: Boston, MA.

Gibson JJ (1979) *The Ecological Approach to Visual Perception*. Houghton-Mifflin: Boston, MA.

Gilbert CD, Wiesel TN (1985) Intrinsic connectivity and receptive field properties in the visual cortex. *Vision Res* 25:365–374.

Gilbert CD, Wiesel TN (1989) Columnar specificity of intrinsic horizontal and corticocortical connections in cat visual cortex. *J Neurosc* 9:2432–2422.

Glauber RJ (1963) Coherent and incoherent states of the radiation field. *Phys Rev* 131:2766–2788.

Gove A, Grossberg S, Mingolla E (1995) Brightness perception, illusory contours, and corticogeniculate feedback. *Visual Neurosci* 12:1027–1052.

Green M (1986) What determines correspondence strength in apparent motion? *Vision Res.* 26:599–607.

Grimson WEL (1982) A computational theory of visual surface interpolation. *Phil Trans Roy Soc Lond B Biol Sci* 298:395–427.

Grimson WEL (1981) *From Images to Surfaces: A Computational Study of the Human Early Visual System*. MIT Press: Cambridge, MA.

Grossberg S, Kuhlmann L, Mingolla E (2007) A neural model of 3D shape-from-texture: multiple-scale filtering, boundary grouping, and surface filling-in. *Vision Res* 47:634–672.

Grossberg S, Mingolla E (1985) Neural dynamics of form perception: Boundary completion, illusory figures, and neon color spreading. *Psych Rev* 92:173–211.

Gu XD, Wang Y, Chan TF, Thompson PM, Yau ST (2004) Genus zero surface conformal mapping and its application to brain surface mapping. *IEEE Trans Med Imaging* 23:940–958.

Gu XD, Luo F, Yau ST (2009) Recent advances in computational conformal geometry. *Commun Inf Syst* 9:163–196.

Gu XD, Yau ST (2008) Computational conformal geometry. *Advanced Lectures in Mathematics*. High Education Press and International Press: Boston, MA.

Guggenheimer HW (1977) *Differential Geometry*. Dover Publications: Mineola, NY.

Haines TS, Wilson RC (2008) Belief propagation with directional statistics for solving the shape-from-shading problem. In *Proc. Eur Conf Comput Vision (ECCV)* 4:780–791.

Hamilton RS (1988) The Ricci flow on surfaces. *Mathematics and General Relativity* (Santa Cruz, CA, 1986), *Contemp Math* 71:237–262.

Hamilton RS (1982) Three manifolds with positive Ricci curvature. *J Diff Geom* 17:255–306.

He ZJ, Nakayama K (1994) Perceived surface shape not features determines correspondence strength in apparent motion. *Vision Res* 34:2125–2135.

Heitger F, Rosenthaler L, von der Heydt R, Peterhans E, Kubler O (1992) Simulation of neural contour mechanisms: From simple to end-stopped cells. *Vision Res* 23:963–981.

Heitger F, von der Heydt R, Peterhans E, Rosenthaler L, Kubler O (1998) Simulation of neural mechanisms: Representing anomalous contours. *Image Vision Comput* 16:407–421.

Heskes T (2004) On the uniqueness of loopy belief propagation fixed points. *Neur Comput* 16:2379–2413.

Heskes T, Albers K, Kappan B (2003) Approimate inference and contrained optimization. In *UAI*, 313–320.

Hess RF, Holliday IE (1992) The coding of spatial position by the human visual system: Effects of spatial scale and contrast. *Vision Res* 32:1085–97.

Hillis JM, Watt SJ, Landy MS, Banks MS (2004) Slant from texture and disparity cues: Optimal cue combination. *J Vision* 4:967–992, doi:10.1167/4.12.1.

Hinkle DA, Connor CE (2002) Three-dimensional orientation tuning in macaque area V4. *Nature Neurosci* 5:665–670.

Hinton G (1999) Products of experts. In *International Conference on Artificial Neural Networks* 1:1–6.

Hoffman WC (1989) The visual cortex is a contact bundle, *Appl Math Comput* 32:137–167.

Horn BKP (1989) Obtaining shape from shading information. In *Shape from shading*. MIT Press: Cambridge, MA, 123–171.

Howe CQ, Purves D (2002) Range image statistics can explain the anomalous perception of length. *Proc Nat Acad Sci* 99:13184–13188.

Huang J, Lee AB, Mumford D (2000) Statistics of range images. In *IEEE Conf Comput Vision Patt Recogn (CVPR)* 16:1324–1331.

Huang T, Russell S (1998) Object identification: A Bayesian analysis with application to traffic surveillance. *Art Intell* 103:1–17.

Huang TS, Lee CH (1989) Motion and structure from orthographic projections. *IEEE Trans Patt Anal Mach Intell (PAMI)* 11:536–540.

Huffman DA (1971) Impossible objects as nonsense sentences. *Mach Intell* 6:295–323.

Huffman DA (1977) Realizable configurations of lines in pictures of polyhedra. *Mach Intell* 8:493–509.

Irani M, Anandan P (1998) A unified approach to moving object detection in 2D and 3D scenes. *IEEE Trans Patt Anal Mach Intell (PAMI)* 20:577–589.

Itti L, Koch C, Niebur E (1998) A model of saliency-based visual attention for rapid scene analysis. *IEEE Trans Patt Anal Mach Intell (PAMI)*, 20:1254–1259.

Ivanchenko VV (2006) *Cue-Dependent and Cue-Invariant Mechanisms in Perception and Perceptual Learning*. PhD Diss, University of Rochester, NY.

Ivanchenko VV, Coughlan J, Gerrey B, Shen H (2008) Computer vision-based clear path guidance for blind wheelchair users. In *Proc 10th Int ACM SIGACCESS Conf Comput Access*, 10:291–292.

Ivanchenko VV, Jacobs RA (2004) Cue-invariant learning for visual slant discrimination. *J Vision* 4:295a.

Ivanchenko VV, Shen H, Coughlan JM (2009) Elevation-based stereo implemented in real-time on a GPU. *IEEE Workshop on Appl Comput Vision*. Snowbird, Utah.

Jacobs RA (2002) What determines visual cue reliability? *Trends Cogn Sci* 6:345–350.

Janssen P, Vogels R, Liu Y, Orban GA (2001) Macaque inferior temporal neurons are selective for three-dimensional boundaries and surfaces. *J Neurosci* 21:9419–9429.

Jin M, Kim J, Luo F, Gu X (2008) Discrete surface Ricci flow. *IEEE Trans Visualiz Comput Graphics (VCG)* 14:1030–1043.

Jin M, Zeng W, Luo F, Gu X (2009) Computing Teichmüller shape space. *IEEE Trans Visualiz Comput Graphics (VCG)* 15:504–517.

Johnston EB, Cumming BG, Parker AJ (1993) Integration of depth modules: Stereopsis and texture. *Vision Res* 33:813–826.

Joly O, Vanduffel W, Orban GA (2009) The monkey ventral premotor cortex processes 3D shape from disparity. *NeuroImage* 47:262–272.

Julesz B (1971) *Foundations of Cyclopean Perception*. University of Chicago Press: Chicago, IL.

Kalar DJ, Garrigan P, Wickens TD, Hilger JD, Kellman PJ (2010) A unified model of illusory and occluded contour interpolation. *Vision Res* 50:284–299.

Kanizsa G (1976) Subjective contours. *Sci Amer* 234:48–52.

Kanizsa G (1979) *Organization in Vision*. Praeger: New York.

Kapadia MK, Gilbert CD, Westheimer G (1995) Improvement in visual sensitivity by changes in local context: Parallel studies in human observers and in V1 of alert monkeys. *Neuron* 15:843–856.

Kass M, Witkin A, Terzopoulos D (1988) Snakes: Active contour models. *Int J Comput Vision* 1:321–331.

Kellman PJ, Garrigan PB, Kalar D, Shipley TF (2003) Good continuation and relatability: Related but distinct principles. *J Vision* 3:120.

Kellman PJ, Garrigan PB, Shipley TF (2005a) Object interpolation in three dimensions. *Psych Rev* 112:586–609.

Kellman PJ, Garrigan PB, Shipley TF, Yin C, Machado L (2005b) 3D interpolation in object perception: Evidence from an objective performance paradigm. *J Exp Psych: Human Percept Perf* 31:558–583.

Kellman PJ, Machado LJ, Shipley TF, Li CC (1996) 3D determinants of object completion. *Invest Ophthal Visual Sci, Supp* 37:S685.

Kellman PJ, Shipley TF (1991) A theory of visual interpolation in object perception. *Cogn Psych* 23:141–221.

Kellman PJ, Yin C, Shipley TF (1998) A common mechanism for illusory and occluded object completion. *Journal J Exp Psych: Human Percept Perf* 24:859–869.

Kharevych L, Springborn B, Schröder P (2006) Discrete conformal mappings via circle patterns. *ACM Trans Graphics* 25:412–438.

Knill DC (2003) Mixture models and the probabilistic structure of depth cues. *Vision Res* 43:831–854.

Knill DC (2007) Robust cue integration: A Bayesian model and evidence from cue-conflict studies with stereoscopic and figure cues to slant. *J Vision* 7:5, 1-24, http://journalofvision.org/7/7/5/, doi:10.1167/7.7.5.

Knill DC (1998) Ideal observer perturbation analysis reveals human strategies for inferring surface orientation from texture. *Vision Res* 38:2635–2656.

Knill DC Saunders JA (2003) Do humans optimally integrate stereo and texture information for judgments of surface slant? *Vision Res* 43:2539–2558.

Koebe P (1936) Kontaktprobleme der Konformen Abbildung. *Ber Sächs Akad Wiss Leipzig, Math.-Phys Kl* 88:141–164.

Koenderink JJ (1984) What does the occluding contour tell us about solid shape? *Perception* 13:321–330.

Koenderink JJ (1990) *Solid Shape*. MIT Press: Cambridge, MA.

Koenderink JJ, van Doorn AJ (1982) The shape of smooth objects and the way contours end. *Perception* 11:129–137.

Koenderink JJ, van Doorn AJ, Kappers AML (1995) Depth relief. *Perception* 24: 115–126.

Koenderink JJ, van Doorn AJ, Kappers AML (1996) Pictorial surface attitude and local depth comparisons. *Percept Psychophys* 58:163–173.

Koenderink JJ, van Doorn AJ, Kappers AML, Todd JT (2001) Ambiguity and the "Mental Eye" in pictorial relief. *Perception,* 30:431–448.

Koffka K (1935) *Principles of Gestalt Psychology*. Routledge & Kegan Paul: London.

Kolmogorov V, Zabih R (2001) Computing visual correspondence with occlusions using graph cuts. *IEEE Int Conf Comput Vision. (ICCV)* 2:508–515.

Kontsevich LL, Tyler CW (1998) How much of the visual object is used in estimating its position? *Vision Res* 38:3025–3029.

Kontsevich LL, Tyler, CW (1999) Bayesian adaptive estimation of psychometric slope and threshold. *Vision Res* 39:2729–2737.

Kovacs I, Julesz B (1993) A closed curve is much more than an incomplete one: Effect of closure in figure–ground segmentation. *Proc Nat Acad Sci* 90:7495–7497.

Kritikos HN, Cho JH (1997) Bargman transform and phase space filters, *Prog Electromag Res* (PIER) 17:45–72.

Kschischang FR, Frey BJ (1998) Iterative decoding of compound codes by probability propagation in graphical models. *IEEE Journal of Selected Areas in Communications (JSAC)* 16:219–230.

Lafferty J, McCallum A, Pereira F (2001) Conditional random fields: Probabilistic models for segmenting and labeling sequence data. In *Proc Int Conf Mach Learning* 18:282–289.

Lan X, Roth S, Huttenlocher DP, Black MJ (2006) Efficient belief propagation with learned higher-order Markov random fields. In *Eur Conf Comput Vision* 10:269–282.

Landy MS, Maloney LT, Johnston EB, Young M (1995) Measurement and modeling of depth cue combination: In defense of weak fusion. *Vision Res* 35:389–412.

Langer MS, Bülthoff HH (1999) Perception of shape from shading on a cloudy day. *Tech Report 73*, Tübingen, Germany.

Langer MS, Zucker SW (1994) Shape from shading on a cloudy day. *J Opt Soc of Amer A* 11:467–478.

Leclerc YG (1989) Constructing simple stable descriptions for image partitioning. *Int J Comput Vision* 3:73–102.

Lee JM, Parker TH (1987) The Yamabe problem. *Bull Amer Math Soc* 17:37–91.

Lee KM, Kuo CCJ (1993) Shape from shading with a linear triangular element surface model. *IEEE Trans Patt Anal Mach Intell (PAMI)* 15:815–822.

Lee TS (1996) Image representation using 2D Gabor wavelets. *IEEE Trans Patt Anal Mach Intell (PAMI)* 18:951–979.

Lee TS, Mumford D (2003) Hierarchical Bayesian inference in the visual cortex. *J Opt Soc Amer A* 20:1434–1448.

Levi DM, Klein SA, Wang H (1994) Discrimination of position and contrast in amblyopic and peripheral vision. *Vision Res* 34:3293–3313.

Levitt JB, Kiper DC, Movshon JA (1994) Receptive fields and functional architecture of macaque V2. *J Neurophysiol* 71:2517–2542.

Lewicki M, Rao RPM, Olshausen BA (eds) (2002) *Probabilistic Models of the Brain: Perception and Neural Function.* Cambridge, MA: MIT Press.

Li FF, Perona P (2005) A Bayesian hierarchical model for learning natural scene categories. *IEEE Comput Vision Patt Recogn (CVPR)* 21:524–531.

Li Y (2009) Perception of parallelepipeds: Perkins's law. *Perception* 38:1767–781.

Li Y, Pizlo Z, Steinman RM (2009) A computational model that recovers the 3D shape of an object from a single 2D retinal representation. *Vision Res* 49:979–991.

Likova LT, Tyler CW (2003) Peak localization of sparsely sampled luminance patterns is based on interpolated 3D object representations. *Vision Res* 43:2649–2657.

Likova LT, Tyler CW (2007) Stereomotion processing in the human occipital cortex. *Neuroimage* 38:293–305.

Liu Y, Vogels R, Orban GA (2004) Convergence of depth from texture and depth from disparity in macaque inferior temporal cortex. *J Neurosci* 24:3795–3800.

Logothetis NK, Pauls J, Poggio T (1995) Shape representation in the inferior temporal cortex of monkeys. *Curr Biol* 5:552–563.

Lowe DG (2004) Distinctive image features from scale-invariant keypoints. *Int J Comput Vision,* 60:91–110.

Lu ZL, Sperling G (1995) The functional architecture of human visual motion perception. *Vision Res* 35:2697–2722.

Lu ZL, Sperling G (2001) Three-systems theory of human visual motion perception: Review and update. *J Opt Soc Amer A* 18:2331–2370.

Luo F (2004) Combinatorial Yamabe flow on surfaces. *Commun Contemp Math* 6:765–780.

Luo F, Gu X, Dai J (2007) *Variational Principles for Discrete Surfaces. Advanced Lectures in Mathematics.* High Education Press and International Press: Boston, MA.

MacCurdy E (ed) (1938) *The Notebooks of Leonardo da Vinci, Volume II.* Reynal & Hitchcock: New York, NY.

Mach E (1906/1959) *The Analysis of Sensations.* Dover: New York.

Mackworth AK (1973) Interpreting pictures of polyhedral scenes. *Artif Intell* 4:121–137.

Malik J (1987) Interpreting line drawings of curved objects. *Int J Comput Vision* 1:73–103.

Malik J, Belongie S, Leung T, Shi JB (2001) Contour and texture analysis for image segmentation. *Int J Comput Vision,* 43:7–27.

Marr D (1982) *Vision: A Computational Investigation into the Human Representation and Processing of Visual Information*. W.H. Freeman and Company: New York.

Martin D, Fowlkes C, Malik J (2004) Learning to detect natural image boundaries using local brightness, color and texture cues. *IEEE Trans Patt Anal Mach Intell (PAMI)* 26:530–549.

Martin D, Fowkes C, TAID, Malik J (2001) A database of human segmented natural images and its application to evaluating segmentation algorithms and measuring ecological statistics. In *Proc 8th Int'l Conf. Computer Vision* 2:416–423.

Maunsell JH, van Essen DC (1983) Functional properties of neurons in middle temporal visual area of the macaque monkey. I. Selectivity for stimulus direction, speed, and orientation. *J Neurophysiol* 49:1127–1147.

Medioni GG, Lee MS, Tang CK (2000) *A Computational Framework for Segmentation and Grouping*. Elsevier: New York.

Mendola JD, Dale AM, Fishl B, Liu AK, Tootell RB (1999) The representation of illusory and real contours in human cortical visual areas revealed by functional magnetic resonance imaging. *J Neurosci* 19:8560–8572.

Mikolajczyk K, Schmid S (2002) An affine invariant interest point detector. In *Proc Eur Conf Comput Vision (ECCV)*. Springer-Verlag.

Miller KD, Kayser A, Priebe NJ (2001) Contrast-dependent nonlinearities arise locally in a model of contrast-invariant orientation tuning. *J Neurophysiol* 85:2130–2149.

Moghaddam B (2001) Principal manifolds and probabilistic subspaces for visual recognition. *IEEE Trans Patt Anal Mach Intell (PAMI)* 24:780–788.

Morgan MJ, Watt RJ (1982) Mechanisms of interpolation in human spatial vision. *Nature* 299:553–555.

Mortensen EN, Barrett WA (1995) Intelligent scissors for image composition. In *SIGGRAPH* 95:191–198.

Mumford D (1994) Elastica and computer vision. In *Algebraic Geometry and its Applications*, West Lafayette, IN, 1990, 491–506. Springer: New York.

Mumford D (1996) Pattern theory: A unifying perspective. In *Perception as Bayesian Inference*, Knill DC, Richards W (eds). Cambridge University Press: Cambridge, UK, 25–62.

Murray MM, Wylie GR, Higgins BA, Javitt DC, Schroeder CE (2002) The spatio-temporal dynamics of illusory contour processing: Combined high-density electrical mapping, source analysis, and functional magnetic resonance imaging. *J Neurosci* 22:5055–5073.

Myers DG (1995) *Psychology*. Worth Publishers: New York.

Nagel A, Stein EM, Wainger S (1985) Balls and metrics defined by vector fields I: Basic properties, *Acta Math* 155:103–147.

Nakayama K, He Z, Shimojo S (1995) Visual surface representation: A critical link between lower-level and higher-level vision. In *An Invitation to Cognitive Science: Visual Cognition*. Osherson DN (ed), 1–70. Cambridge, MA: The MIT Press.

Nakayama K, Shimojo S (1990) Towards a neural understanding of visual surface representation. *The Brain: Cold Spring Harbor Symposium on Quantitative Biology,* In Sejnowski T, Kandel ER, Stevens CF, Watson JD (eds), 55:911–924. Cold Spring Harbor Laboratory: Cold Spring Harbor, NY.

Nakayama K, Shimojo S (1996) Experiencing and perceiving visual surfaces. In *Perception as Bayesian Inference,* Knill DC, Richards W (eds), 391–407. Cambridge, MA: Cambridge University Press.

Nakayama K, Shimojo S, Silverman GH (1989) Stereoscopic depth: Its relation to image segmentation, grouping, and the recognition of occluded objects. *Perception* 18:55–68.

Nayar SK, Narasimhan SG (1999) Vision in bad weather. In *IEEE Int Conf Comput Vision (ICCV)* 2:820–827.

Neisser U (1967) *Cognitive Psychology.* Appleton-Century-Crofts: East Norwalk, CT.

Nelson RC (1991) Qualitative detection of motion by a moving observer. *Int J Comput Vision,* 7:33–46.

Nelson SB, Sur M, Somers DC (1995) An emergent model of orientation selectivity in cat visual cortical simple cells. *J Neurosci* 15:5448–5465.

Nguyenkim JD, DeAngelis GC (2003) Disparity-based coding of three-dimensional surface orientation by macaque middle temporal neurons. *J Neurosci* 23:7117–7128.

Nienborg H, Bridge H, Parker AJ, Cumming BG (2004) Receptive field size in V1 neurons limits acuity for perceiving disparity modulation. *J Neurosci* 24:2065–2076.

Nienborg H, Bridge H, Parker AJ, Cumming BG (2005) Neuronal computation of disparity in V1 limits temporal resolution for detecting disparity modulation. *J Neurosci* 25:10207–10219.

Nitzberg M, Mumford D, Shiota T (1993) *Filtering, Segmentation and Depth.* Springer: Berlin.

Norman JF, Todd JT (1996) The discriminability of local surface structure. *Perception* 25:381–398.

Norman JF, Todd JT (1998) Stereoscopic discrimination of interval and ordinal depth relations on smooth surfaces and in empty space. *Perception* 27:257–272.

Norman JF, Todd JT, Phillips F (1995) The perception of surface orientation from multiple sources of optical information. *Percept Psychophys* 57:629–636.

Odobez JM, Bouthemy P (1995) MRF-based motion segmentation exploiting a 2D motion model robust estimation. In *Int Conf Image Process* 95:3628.

Ott T, Stoop R (2006) The neurodynamics of belief propagation on binary Markov random fields. *Neur Inf Process Syst (NIPS)* 19:1082–1088.

Pahlavan, K, Uhlin T, Eklundh JO (1996) Dynamic fixation and active perception. *Int J Comput Vision,* 17:113–135.

Palmer EM, Kellman PJ, Shipley TF (2006) A theory of dynamic occluded and illusory object perception. *J Exp Psych: Gen* 35:513–541.

Patterson R (1999) Stereoscopic (cyclopean) motion sensing. *Vision Res* 39: 3329–3345.

Patterson R (2002) Three-systems theory of human visual motion perception: Review and update: comment. *J Opt Soc Am A* 19:2142–2143.

Pearl J (1998) *Probabilistic Reasoning in Intelligent Systems: Networks of Plausible Inference*. Morgan Kaufman: San Francisco, CA.

Pentland AP (1990) Linear shape from shading. *International Journal of Computer Vision* 4:153–162, March.

Perelman G (2002) The entropy formula for the Ricci flow and its geometric applications. *Technical Report arXiv.org*, Nov 11 2002.

Perelman G (2003a) Finite extinction time for the solutions to the Ricci flow on certain three-manifolds. *Technical Report arXiv.org*, Jul 17 2003.

Perelman G (2003b) Ricci flow with surgery on three-manifolds. *Technical Report arXiv.org*, Mar 10 2003.

Perelomov AM (1986) *Generalized Coherent States and their Applications*. Springer-Verlag: Berlin.

Peterhans E, von der Heydt R (1989) Mechanisms of contour perception in monkey visual cortex. II. Contours bridging gaps. *J Neurosci* 9:1749–1763.

Petitot J (1994) Phenomenology of perception, qualitative physics and sheaf mereology, in *Proceedings of the 16th International Wittgenstein Symposium, Vienna*, Verlag Hlder-Pichler-Tempsky, 387–408.

Petitot J, Tondut Y (1999) Vers une Neurogeometrie. Fibrations corticales, structures de contact et contours subjectifs modaux. *Mathematiques, Informatique et Sciences Humaines*, 145, EHESS, CAMS, Paris, 5–101.

Pizlo Z (2001) Perception viewed as an inverse problem. *Vision Res* 41:3145–3161.

Pizlo Z (2008) *3D Shape: Its Unique Place in Visual Perception*. MIT Press: Cambridge.

Pizlo Z, Sawada T, Li Y, Kropatsch WG, Steinman RM (2010) New approach to the perception of 3D shape based on veridicality, complexity, symmetry and volume. *Vision Res* 50:1–11.

Poggio T, Torre V, Koch C (1985) Computational vision and regularization theory. *Nature* 317:314–319.

Polat U, Sagi D (1994) The architecture of perceptual spatial interactions. *Vision Res* 34:73–78.

Pomerantz JR, Kubovy M (1986) Theoretical approaches to perceptual organization: Simplicity and likelihood principles. In *Handbook of Perception and Human Performance: Vol. 2. Cognitive Processes and Performance*, Boff KR, Kaufman L, Thomas JP (eds). Wiley: New York, 36.1–36.46.

Potetz B (2007) Efficient belief propagation for vision using linear constraint nodes. In *CVPR 2007: Proceedings of the 2007 IEEE Computer Society Conference on Computer Vision and Pattern Recognition*. IEEE Computer Society, Minneapolis, MN.

Potetz B, Lee TS. (2003) Statistical correlations between two-dimensional images and three-dimensional structures in natural scenes. *J Opt Soc Amer A* 20:1292–1303.

Potetz B, Lee TS (2006) Scaling laws in natural scenes and the inference of 3d shape. In Weiss Y, Schölkopf B, Platt J (eds), *Advances in Neural Information Processing Systems* 18:1089–1096. Cambridge, MA: MIT Press.

Potetz B, Lee TS (2008) Efficient belief propagation for higher order cliques using linear constraint nodes. *Comput Vision Image Understand* 112:39–54.

Priebe NJ, Miller KD, Troyer TW, Krukowsky AE (1998) Contrast-invariant orientation tuning in cat visual cortex: thalamocortical input tuning and correlation-based intracortical connectivity. *J Neurosci* 18:5908–5927.

Qiu FT, Sugihara T, von der Heydt R (2007) Figure–ground mechanisms provide structure for selective attention. *Nature Neurosci* 10:1492–1499.

Qiu FT, von der Heydt R (2005) Figure and ground in the visual cortex: V2 combines stereoscopic cues with Gestalt rules. *Neuron* 47:155–166.

Qiu FT, von der Heydt R (2007) Neural representation of transparent overlay. *Nature Neurosci* 10: 283–284.

Ramachandran VS, Ruskin D, Cobb S, Rogers-Ramachandran D, Tyler CW (1994) On the perception of illusory contours. *Vision Res* 34:3145–352.

Rao RPM (2004) Bayesian computation in recurrent neural circuits. *Neural Comput* 16:1–38.

Regan D, Hajdur LV, Hong XH (1996) Two-dimensional aspect ratio discrimination for shape defined by orientation texture. *Vision Res* 36:3695–3702.

Regan D, Hamstra SJ (1991) Shape discrimination for motion-defined and contrast-defined form: squareness in special. *Perception* 20:315–336.

Regan D, Hamstra SJ (1994) Shape discrimination for rectangles defined by disparity alone, by disparity plus luminance and by disparity plus motion. *Vision Res* 34:2277–2291.

Ren X, Malik J (2002) A probabilistic multi-scale model for contour completion based on image statistics. In *Proc Eur Conf Comput Vision (ECCV)* 7:312–327.

Ringach DL, Shapley R (1996) Spatial and temporal properties of illusory contours and amodal boundary completion. *Vision Res* 36:3037–3050.

Rizzolatti G, Fadiga L, Gallese V, Fogassi L (1996) Premotor cortex and the recognition of motor actions. *Cogn Brain Res* 3:131–141.

Rizzolatti G, Sinigaglia C (2010) *Mirrors in the Brain: How Our Minds Share Actions, Emotions, and Experience.* Oxford University Press: Oxford.

Rodin B, Sullivan D (1987) The convergence of circle packings to the Riemann mapping. *J Diff Geom* 26:349–360.

Rollenhagen JE, Olson CR (2000) Mirror-image confusion in single neurons of the macaque inferotemporal cortex. *Science* 287:1506–1508.

Roth S, Black MJ (2005) Fields of experts: A framework for learning image priors. In *CVPR*, 860–867.

Rother C, Kolmogorov V, Blake A (2004) "Grabcut": Interactive foreground extraction using iterated graph cuts. *ACM Trans Graph* 23:309–314.

Rubin E (1921) *Visuell wahrgenommene Figuren.* Gyldendals: Copenhagen.

Rubin N (2001) The role of junctions in surface completion and contour matching. *Perception* 30:339–366.

Ruderman DL, Bialek W (1994) Statistics of natural images: Scaling in the woods. *Phys Rev Lett* 73:814–817.

Rue H, Hurn MA (1999) Bayesian object identification. *Biometrika* 86:649–660.

Sabra AI (1989) (ed and transl) *The Optics of Ibn Al-Haytham.* The Warburg Institute: London, 155.

Saidpour A, Braunstein ML, Hoffman DD (1994) Interpolation across surface discontinuities in structure from motion. *Percept Psychophys* 55:611–622.

Sajda P, Finkel LH (1995) Intermediate-level visual representations and the construction of surface perception. *J Cogn Neurosci* 7:267–291.

Sakata H, Taira M, Kusunoki M, Murata A, Tanaka Y (1997) The parietal association cortex in depth perception and visual control of hand action. *Trends Neurosci* 20:350–357.

Sakata H, Taira M, Kusunoki M, Murata A, Tsutsui K, Tanaka Y, Shein WN, Miyashita Y (1999) Neural representation of three-dimensional features of manipulation objects with stereopsis. *Exp Brain Res* 128:160–169.

Samonds JM, Potetz BR, Lee TS (2009) Cooperative and competitive interactions facilitate stereo computations in macaque primary visual cortex. *J Neurosci* 29:15780–15795.

Sarti A, Malladi R, Sethian JA (2000) Subjective surfaces: A method for completing missing boundaries. *Proc Nat Acad Sci USA* 12:6258–6263.

Sarti S, Citti G (2010) A stochastic model for edges in natural images using Lie groups. *J Vision* (in press).

Sarti S, Citti G, Petitot J (2008) The symplectic structure of the primary visual cortex. *Biol Cybern* 98:33–48.

Sawada T (2010) Visual detection of symmetry of 3D shapes. J Vision, 10:4. doi: 10.1167/10.6.4.

Sawada T, Pizlo Z (2008a) Detecting mirror-symmetry of a volumetric shape from its single 2D image. *IEEE CVPR Workshop Percept Org Comput Vision*, 6:1–8.

Sawada T, Pizlo Z (2008b) Detection of skewed symmetry. *J Vision* 8:14:1–18. doi: 10.1167/8.5.14

Sawhney HS, Guo Y, Kumar R (2000) Independent motion detection in 3D scenes. *IEEE Trans Patt Anal Mach Intell (PAMI)*, 22:1191–1199.

Saxena A, Sun M, Ng A (2007) Learning 3D scene structure from a single still image. In *ICCV Workshop 3D Represent Recogn (3D RR-07)*.

Scharstein D, Pal C (2005) Learning conditional random fields for stereo. In *IEEE Proc Comput Vision Patt Recogn (CVPR)* 21:838–845.

Scharstein D, Szeliski R (2002) A taxonomy and evaluation of dense two-frame stereo correspondence algorithms. *Int J Comput Vision* 47:7–42.

Schoen R (1984) Conformal deformation of a Riemannian metric to constant scalar curvature. *J Diff Geom* 20:479–495.

Seghier M, Dojat M, Delon-Martin C, Rubin C, Warnking J, Segebarth C, Bullier J (2000) Moving illusory contours activate primary visual cortex: An fMRI study. *Cereb Cortex* 10:663–670.

Seitz SM, B Curless B, Diebel J, Scharstein D, Szeliski R (2006) A comparison and evaluation of multi-view stereo reconstruction algorithms. In *IEEE Proc Comput Vision Patt Recogn (CVPR)* 22:519–526.

Serences JT, Yantis S (2006) Selective visual attention and perceptual coherence. *Trends Cogn Sci*, 10:38–45.

Shahar B, Zucker S (2003) Geometrical computations explain projection patterns of long-range horizontal connections in visual cortex. *Neural Comput* 16:445–476.

Sheffield TM, Meyer D, Payne B, Lees J, Harvey EL, Zeitlin MJ, Kahle G (2000). Geovolume visualization interpretation: A lexicon of basic techniques. *The Leading Edge* 19:518–525.

Shelley M, Wielaard DJ, McLaughlin D, Shapley R (2000) A neuronal network model of macaque primary visual cortex (V1): Orientation selectivity and dynamics in the input layer 4Cα. *Proc Nat Acad Sci USA* 97:8087–8092.

Shi J, Malik J (2000) Normalized cuts and image segmentation. *IEEE Trans Patt Anal Mach Intell (PAMI)* 22:888–905.

Shikata E, Tanaka Y, Nakamura H, Taira M, Sakata H (1996) Selectivity of the parietal visual neurones in 3D orientation of surface of stereoscopic stimuli. *Neuroreport* 7:2389–2394.

Shipley TF, Kellman PJ (1990) The role of discontinuities in the perception of subjective figures. *Percept Psychophys* 48:259–270.

Shipley TF, Kellman PJ (1992a) Perception of partly occluded objects and illusory figures. *J Exp Psych: Human Percept Perf* 18:106–120.

Shipley TF, Kellman PJ (1992b) Strength of visual interpolation depends on the ratio of physically specified to total edge length. *Percept Psychophys* 52:97–106.

Sinclair D (1993) Motion segmentation and local structure. *Int Conf Comput Vision* 93:366–373.

Singh M, Hoffman DD (1999a) Completing visual contours: The relationship between relatability and minimizing inflections. *Percept Psychophys* 61:943–951.

Singh M, Hoffman DD (1999b) Contour completion and relative depth: Petter's rule and support ratio. *Psych Sci* 10:423–428.

Sinop AK, Grady L (2007) A seeded image segmentation framework unifying graph cuts and random walker which yields a new algorithm. In *Int Conf Comput Vision* 11:1–8.

Sperling G (1960) The information available in brief visual presentation. *Psych Monog* 74:29.

Springborn B, Schröder P, Pinkall U (2008) Conformal equivalence of triangle meshes. *ACM Trans Graphics* 27:1–11.

Srivastava S, Orban GA, De Mazière PA, Janssen P (2009) A distinct representation of three-dimensional shape in macaque anterior intraparietal area: Fast, metric, and coarse. *J Neurosci* 29:10613–10626.

Stormont DP (2007) An online Bayesian classifier for object identification. *IEEE Int Workshop on Safety, Security and Rescue Robotics* 1–5.

Sudarshan ECG (1963) Equivalence of semiclassical and quantum mechanical descriptions of statistical light beams, *Phys Rev Lett* 10:277–279.

Sugihara K (1986). Machine interpretator of live drawings. Cambridge: MIT Press.

Sugihara H, Murakami I, Shenoy KV, Andersen RA, Komatsu H (2002) Response of MSTd neurons to simulated 3D orientation of rotating planes. *J Neurophysiol* 87:273–285.

Sugita Y (1999) Grouping of image fragments in primary visual cortex. *Nature* 401:269–272.

Sullivan J, Blake A, Isard M, MacCormick J (2001) Bayesian object localisation in images. *Proc Int J Comput Vision* 44:111–135.

Sun J, Shum HY, Zheng NN (2002) Stereo matching using belief propagation. In *Computer Vision: ECCV, Lecture Notes in Computer Science*, Heyden A, Sparr G, Nielsen M, Johansen P (eds). 2351:510–524.

Sun J, Zheng NN, Shum, HY (2003) Stereo matching using belief propagation. *IEEE Trans Pattern Anal Mach Intell (PAMI)* 25:787–800.

Surdick RT, Davis ET, King RA, Hodges LF (1997) The perception of distance in simulated visual displays: A comparison of the effectiveness and accuracy of multiple depth cues across viewing distances. *Presence: Teleoperators Virtual Env* 6:513–531.

Taira M, Tsutsui KI, Jiang M, Yara K, Sakata H (2000) Parietal neurons represent surface orientation from the gradient of binocular disparity. *J Neurophysiol* 83:3140–3146.

Tanaka H, Uka T, Yoshiyama K, Kato M, Fujita I (2001) Processing of shape defined by disparity in monkey inferior temporal cortex. *J Neurophysiol* 85:735–744.

Tang KL, Tang CK, Wong TT (2005) Dense photometric stereo using tensorial belief propagation. In *IEEE Conf Comput Vision Patt Recogn (CVPR)* 21:132–139.

Tarr MJ, Kriegman DJ (2001) What defines a view? *Vision Res* 41:1981–2004.

Taylor IL, Sumner FC (1945) Actual brightness and distance of individual colors when their apparent distance is held constant. *J Psychol* 19:79–85.

Thompson WB, Pong TC (1990) Detecting moving objects. *Int J Comput Vision* 4:39–57.

Thurston WP (1980) *Geometry and Topology of Three-Manifolds.* Princeton University Lecture Notes. Princeton University Press: Princeton, NJ.

Thurston WP (1985) The finite Riemann mapping theorem. Talk at *An International Symposium at Purdue University on the Occasion of the Proof of the Bieberbach Conjecture,* March, 1985.

Todd JT, Bressan P (1990) The perception of 3-dimensional affine structure from minimal apparent motion sequences. *Percept Psychophys* 48:419–430.

Todd JT, Chen L, Norman JF (1998) On the relative salience of Euclidean, affine and topological structure for 3D form discrimination. *Perception* 27:273–282.

Todd JT, Norman JF (1995) The visual discrimination of relative surface orientation. *Perception* 24:855–866.

Todd JT, Norman JF, Koenderink JJ, Kappers AML (1997) Effects of texture, illumination and surface reflectance on stereoscopic shape perception. *Perception* 26:806–822.

Todd JT, Reichel FD (1989) Ordinal structure in the visual perception and cognition of smoothly curved surfaces. *Psych Rev* 96:643–657.

Toet A, Koenderink JJ (1988) Differential spatial displacement discrimination thresholds for Gabor patches. *Vision Res* 28:133–143.

Tola E, Lepetit V, Fua P (2008) A fast local descriptor for dense matching. In *Proc Comput Vision Patt Recogn (CVPR)* 28:1–8.

Tondut Y, Petitot J (1999) Vers une Neurogeometrie. Fibrations corticales, structures de contact et contours subjectifs modaux. *Math Inf Sci Hum, EHESS, CAMS, Paris* 145:5–101.

Torr PHS, Faugeras O, Kanade T, Hollinghurst N, Lasenby J, Sabin M, Fitzgibbon A (1998) Geometric motion segmentation and model selection [and discussion]. *Phil Trans Math Phys Eng Sci* 356:1321–1340.

Torr PHS, Murray DW (1994) Stochastic motion clustering. In *Eur Conf Comput Vision (ECCV)* 3:328–337.

Torralba A, Freeman WT (2003) Properties and applications of shape recipes. In *IEEE Conf Comput Vision Patt Recogn (CVPR)* 23:383–390.

Torralba A, Murphy KP, Freeman WT, Rubin MA (2003) Context-based vision system for place and object recognition. In *IEEE Intl Conf Computer Vision (ICCV)* 53:169–191.

Torralba A, Oliva A (2002) Depth estimation from image structure. *IEEE Trans Pattern Anal Mach Intell (PAMI)* 24:1226–1238.

Triggs W, McLauchlan PF, Hartley RI, Fitzgibbon AW (2000) Bundle adjustment —A modern synthesis. In *IEEE Intl Conf Computer Vision (ICCV): Proc Int Workshop Vision Alg* 1:298–372.

Trudinger NS (1968) Remarks concerning the conformal deformation of Riemannian structures on compact manifolds. *Ann Scuola Norm Sup Pisa* 22:265–274.

Tse PU (1999) Volume completion. *Cognit Psychol* 39:37–68.

Tsutsui K, Jiang M, Yara K, Sakata H, Taira M (2001) Integration of perspective and disparity cues in surface-orientation-selective neurons of area CIP. *J Neurophysiol* 86:2856–2867.

Tsutsui K, Sakata H, Naganuma T, Taira M (2002) Neural correlates for perception of 3D surface orientation from texture gradient. *Science* 298:409–412.

Tu ZW, Zhu SC (2002a) Image segmentation by data-driven Markov chain Monte Carlo. *IEEE Trans Pattern Anal Mach Intell (PAMI)* 24:657–673.

Tu ZW, Zhu SC (2002b) Mean shift: A robust approach toward feature space analysis. *IEEE Trans Pattern Anal Mach Intell (PAMI)* 24:603–619.

Tyler CW (1973) Stereoscopic vision: Cortical limitations and a disparity scaling effect. *Science* 181:276–278.

Tyler CW (1991) Cyclopean vision. In: *Vision and Visual Dysfunction; Vol 9: Binocular Vision*. Regan D (ed), MacMillan: London, 38–74.

Tyler CW (1998) Diffuse illumination as a default assumption for shape from shading in the absence of shadows. *J Imag Sci Tech* 42:319–325.

Tyler CW (2004) Theory of texture discrimination based on higher-order perturbation in individual texture samples. *Vision Res* 44:2179–2186.

Tyler CW (2006) Spatial form as inherently three-dimensional. In *Seeing Spatial Form*, Jenkin MRM and Harris LR (eds). Oxford University Press: Oxford, 67–88.

Tyler CW (2010) Introduction: The role of mid-level surface representation in 3D object encoding. In *Computer Vision: From Surfaces to 3D Objects*. Tyler CW (ed), Taylor & Francis: London.

Tyler CW, Kontsevich LL (1995) Mechanisms of stereoscopic processing: Stereo-attention and surface perception in depth reconstruction. *Perception* 24: 127–153.

Tyler CW, Likova LT, Kontsevich LL, Wade AR (2006) The specificity of cortical area KO to depth structure. *NeuroImage* 30:228–238.

Ullman S (1976) Filling in the gaps: The shape of subjective contours and a model for their generation. *Biol Cybernet* 25:1–6.

van Ee R, Adams WJ, Mamassian P (2003) Bayesian modeling of cue interaction: Bistability in stereoscopic slant perception. *J Opt Soc Am A* 20:1398–1406.

van Hateren JH, van der Schaaf A (1998) Independent component filters of natural images compared with simple cells in primary visual cortex. *Proc Roy Soc B* 265:359–366.

Van Oostende S, Sunaert S, Van Hecke P, Marchal G, Orban GA (1997) The kinetic occipital (KO) region in man: An fMRI study. *Cereb Cortex* 7:690–701.

Veksler O (2008) Star shape prior for graph-cut image segmentation. In *Eur Conf Comput Vision (ECCV)* 10:454–467.

Vetter T, Poggio T (1994) Symmetric 3D objects are an easy case for 2D object recognition. *Spat Vision* 8:443–453.

Vetter T, Poggio T, Bülthoff HH (1994) The importance of symmetry and virtual views in three-dimensional object recognition. *Curr Biol* 4:18–23.

von der Heydt R, Friedman HS, Zhou H (2003) Searching for the neural mechanisms of color filling-in. In: *Filling-in: From Perceptual Completion to Cortical Reorganization.* Pessoa L, De Weerd P (eds). Oxford University Press: Oxford, 106–127.

von der Heydt R, Zhou H, Friedman HS (2000) Representation of stereoscopic edges in monkey visual cortex. *Vision Res* 40:1955–1967.

Wallach H, O'Connell DN (1953) The kinetic depth effect. *J Exp Psychol* 45:205–217.

Wallach H, Zuckerman C (1963) The constancy of stereoscopic depth. *Amer J Psychol*, 76:404–412.

Wallschlaeger C, Busic-Snyder C (1992) *Basic Visual Concepts and Principles for Artists, Architects, and Designers.* McGraw Hill: Boston.

Walther D, Koch C (2006) Modeling attention to salient proto-objects. *Neural Networks*, 19:1395–1407.

Waltz D (1975) Generating semantic descriptions from drawings of scenes with shadows. In *The Psychology of Computer Vision*, Winston PH (ed), McGraw-Hill: New York, 19–91.

Weber J, Malik J (1997) Rigid body segmentation and shape description from dense optical flow under weak perspective. *IEEE Trans Patt Anal Mach Intell (PAMI)* 19:139–143.

Weiss Y (1997a) Interpreting images by propagating Bayesian beliefs. In *Adv Neur Inf Proc Syst (NIPS)* 9:908–915.

Weiss Y (1997b) Smoothness in layers: Motion segmentation using nonparametric mixture estimation. *IEEE Conf Comput Vision Patt Recogn (CVPR)* 13:520–526.

Weiss Y, Freeman WT (2001) On the optimality of solutions of the max-product belief propagation algorithm in arbitrary graphs. *IEEE Trans Inf Theory, Special Issue on Codes on Graphs and Iterative Algorithms* 47:723–735.

Weiss Y, Freeman WT (2007) What makes a good model of natural images? In *CVPR 2007: Proceedings of the 2007 IEEE Computer Society Conference on Computer Vision and Pattern Recognition*. Minneapolis, MN: IEEE Computer Society.

Weitraub SH (2007) *Differential Forms: A Complement to Vector Calculus.* Academic Press: New York.

Wertheimer M (1923) Untersuchungen zur lehre von der gestalt II [Laws of organization in perceptual forms]. *Psyc Forsch* 4:301–350.

Wertheimer M (1923/1938) Principles of perceptual organization. In *Readings in Perception*, Beardslee DC, Wertheimer M (eds), D. Van Nostrand: New York, 115–135.

White R, Forsyth D (2006) Combining cues: Shape from shading and texture. In *Proc Comput Vision Patt Recogn (CVPR)* 22:1809–1816.

Wiles CS, Brady M (1995). Closing the loop on multiple motions. In *Proc Int Conf Comput Vision (ICCV)* 5:308–313.

Williams LR, Jacobs DW (1997a) Local parallel computation of stochastic completion fields. *Neural Comput* 9:859–881.

Williams LR, Jacobs DW (1997b) Stochastic completion fields: a neural model of illusory contour shape and salience. *Neural Comput* 9:837–858.

Winston PH (1975) Learning structural descriptions from examples. In *The Psychology of Computer Vision*, Winston PH (ed), McGraw-Hill: New York, 157–200.

Woodford OJ, Reid ID, Torr PHS, Fitzgibbon, AW (2006) Fields of experts for image-based rendering. In *Proceedings of the 17th British Machine Vision Conference*, Edinburgh 3:1109–1108.

Woodford OJ, Torr PHS, Reid ID, Fitzgibbon AW (2008) Global stereo reconstruction under second order smoothness priors. In *IEEE Proc Comput Vision Patt Recogn (CVPR)* 24:2115–2128.

Wright M, Ledgeway T (2004) Interaction between luminance gratings and disparity gratings. *Spatial Vision* 17:51–74.

Xu C, Prince JL (1998) Snakes, shapes, and gradient vector flow. *IEEE Trans Image Process* 7:359–369.

Yamabe H (1960) The Yamabe problem. *Osaka Math J* 12:21–37.

Yang Z, Purves D (2003) Image/source statistics of surfaces in natural scenes. *Network* 14:371–390.

Yarbus AL (1967) *Eye Movements and Vision.* Plenum Press: NY.

Yedidia JS, Freeman WT, Weiss Y (2000) Generalized belief propagation. In *Adv Neur Inf Proc Syst (NIPS)* 12:689–695.

Yedidia JS, Freeman WT, Weiss Y (2001) Understanding belief propagation and its generalizations. *MERL Cambridge Research Tech Report* TR 2001–2016.

Yedidia JS, Freeman WT, Weiss Y (2003) Understanding belief propagation and its generalizations. In Lakemeyer G, Nebel B (eds), *Exploring Artificial Intelligence in the New Millennium*. Morgan Kaufmann: San Francisco, 239–269.

Yedidia JS, Freeman WT, Weiss Y (2005) Constructing free-energy approximations and generalized belief propagation algorithms. *IEEE Trans Inf Theory* 51:2282–2312.

Yen SC, Finkel LH (1998) Extraction of perceptually salient contours by striate cortical networks. *Vision Res 38*:719–741.

Yin C, Kellman PJ, Shipley TF (1997) Surface completion complements boundary interpolation in the visual integration of partly occluded objects. *Perception 26*:1459–1479.

Yin C, Kellman PJ, Shipley TF (2000) Surface integration influences depth discrimination. *Vision Res 40*:1969–1978.

Young MJ, Landy MS, Maloney LT (1993) A perturbation analysis of depth perception from combinations of texture and motion cues. *Vision Res* 33:2685–2696.

Yu SX, Shi J (2001) Grouping with bias. In *Adv Neur Inf Process Syst (NIPS)* 13:1327–1334.

Yuille AL (2002) CCCP algorithms to minimize the Bethe and Kikuchi free energies: Convergent alternatives to belief propagation. *Neural Comput* 14:1691–1722.

Yves C de V (1991) Un principe variationnel pour les empilements de cercles. *Invent Math* 104:655–669.

Zahn CT (1971) Graph-theoretical methods for detecting and describing gestalt clusters. *IEEE Trans Comput* 20:68–86.

Zemel R, Behrmann M, Mozer M, Bavelier D (2002) Experience-dependent perceptual grouping and object-based attention. *J Exp Psych* 28:202–217.

Zeng W, Zeng Y, Wang Y, Gu X, Samaras D (2008) 3D non-rigid surface matching and registration based on holomorphic differentials. In *Eur Soc Comput Vision (ECCV)* 10:1–14.

Zhang R, Tsai PS, Cryer JE, Shah M (1999) Shape from shading: A survey. *IEEE Trans Pattern Anal Mach Intell (PAMI)* 21:690–706.

Zhang J (2005) Object oneness: the essence of the topological approach to perception. *Visual Cogn* 12:683–690.

Zhang J, Wu S (1990). Structure of visual perception. *Proc Nat Acad Sci, USA* 87:7819–7823.

Zhang L, Seitz S (2005) Parameter estimation for MRF stereo. In *IEEE Proc Comput Vision Patt Recogn (CVPR)* 21:288–295.

Zheng Q, Chellappa R (1991) Estimation of illuminant direction, albedo, and shape from shading. *IEEE Trans Pattern Anal Mach Intell (PAMI)* 13:680–702.

Zhang Z, Faugeras OD, Ayache N (1988) Analysis of a sequence of stereo scenes containing multiple moving objects using rigidity constraints. In *Int J Comput Vision*, 2:177–186.

Zheng Q, Chellapa R (1993) Motion detection in image sequences acquired from a moving platform. In *Int Conf Acoust, Speech, Signal Process*, 18:201–204.

Zhou H, Friedman HS, von der Heydt R (2000) Coding of border ownership in monkey visual cortex. *J Neurosci* 20:6594–6611.

Zhu SC, Tu ZW (2002) Image segmentation by data-driven Markov chain Monte Carlo. *IEEE Trans Patt Anal Mach Intell (PAMI)* 24:657–673.

Zhu SC, Wu YN, Mumford D (1998) Frame : Filters, random fields and maximum entropy—towards a unitied theory for texture modeling. *Int J Comput Vision* 27:1–20.

Zimmerman GL, Legge GE, Cavanagh P (1995) Pictorial depth cues: A new slant. *J Opt Soc Am A* 12:17–26.

Zucker SW (2000) The curve indicator random field: Curve organization via edge correlation. In *Perceptual Organization for Artificial Vision Systems*, Boyer KL, Sarkar S (eds). Kluwer Academic: Norwell MA, 265–288.

# Index

Printed and bound by CPI Group (UK) Ltd, Croydon, CR0 4YY

23/10/2024

01778266-0002